SPACECRAFT EGRESS AND RESCUE OPERATIONS

Planning for and Managing Post-Landing Contingencies in Crewed Space Missions

Jason D. Reimuller, Ph.D.

Published By:

Integrated Spaceflight Services

Integrated Spaceflight Publications Office
Boulder, Colorado

Integrated Spaceflight Publications Office
3360 Mitchell Lane, Suite C
Boulder, Colorado 80301

First Printing 2015

ISBN: 978-0-9971472-0-9

Integrated Spaceflight Services Website: www.integratedspaceflight.com

Printed and bound in the United States of America.

CONTENTS

Acknowledgements

The author would like to express his gratitude to all the talented engineers with whom he had worked with at NASA during the development of the Constellation Program. In particular, Dr. Jay Falker, Tom Walker, Alan Rhodes, Don Pearson, Dave Forrest, Tim Propp, Jim Hamblin, Don Hammel, Young Lee, and Rick Banke.

The author would also like to acknowledge the contributions of Dr. Erik Seedhouse and Brian Shiro for providing their expertise on commercial scientist-astronaut training and of Maria Hanna and Survival Systems USA for providing their expertise on egress operations as they pertain to air crews.

The contributions of Bruce Hulley, Joe Tanner, Dr. David Klaus, Dr. Waleed Abdalati, and Dr. Stephen Corda have provided the author invaluable experiences and insight to the operational world of spaceflight, and this book could not have been written without their support.

The author would also like to acknowledge the contributions of the space system and space architecture engineers that have worked to make space travel safer. In particular, the author would like to honor the crews of Apollo 1, Soyuz 1, and Soyuz 11 whose tragic deaths led the way to the development of safer landing and egress systems: Roger B. Chaffee Georgi Dobrovolsky, Virgil "Gus" Grissom, Sergei Korolev, Viktor Patsayev, Vladislav Volkov, and Edward H. White II.

Finally, to my parents, Particia Chapman and David Reimuller for their unwavering support, my grandfather B/Gen Willis Chapman and my uncle David Middleton for their guidance towards my career in crewed spaceflight.

This book is dedicated Mr. Don Hammel (1964-2013)

an exceptional post-landing space architect,
NASA mission operator,
and friend.

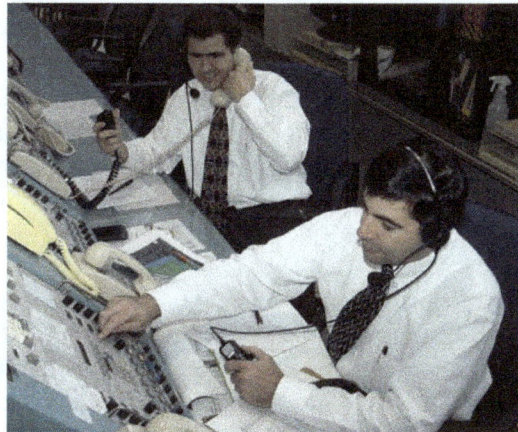

Preface

One third of crewed spaceflights suffer major problems that threaten completion of the mission and the life of the astronauts. Five crews, comprising 2% of all crewed missions, have perished in their spacecraft; spaceflight is anything but routine. In the design of space missions, there are always contingency mission profiles designed to save the crew and hopefully the spacecraft even if the mission is lost. These contingency mission profiles do not necessarily lead to a landing in the location originally planned. Further, problems that can occur during re-entry and descent can lead to landing in an unplanned location. Such landings may drive the probability that the crew would not safely be able to remain in the spacecraft until rescue teams arrive. Even if the landing was nominal, there are always failures in the post-landing phase (the phase between landing and rescue of the crew by support teams) that may force the crew to egress the spacecraft unassisted. Emergency egress operations are one of the most important skills flight crews are trained to perform in preparation for a space mission.

This book presents the evolution of the planning associated with crew emergency egress and rescue operations and then presents an overview of the types of egress systems and procedures that have historically been used on crewed space. An understanding of pre-launch and post-landing failures that can lead to egress operations, as with the hazardous environments that these failures might create, is investigated. Further, the uncertainties of the human factors involved with egress operations are assessed, as with managing the effects of deconditioning from long-duration spaceflight and injury potential.

This book was conceived as the first group of commercial crewed spaceflight companies was developing their own disparate designs that would provide access to suborbital and orbital space. The author's experience as a system engineer for NASA's Constellation Program provided many learned lessons and an understanding of both effective methods to address multi-faceted issues complicated by human-factors issues, the randomness of environments spacecraft may end up in, and the international complexities of contingency rescue plans. The same experience also provided sensitivity to the pitfalls that could be encountered. It is the author's intention to preserve the maturation of knowledge associated with the contingency planning of space missions in order to better assist commercial space architecture developers and future operators.

In this manner, the book addresses essential design trades in crewed spaceflight architecture design, filling a need of the engineers and designers of these systems, but it also is written to serve the needs of future operators of these vehicles. The needs of spaceflight architecture developers and the needs of commercial space vehicle operators are different but complementary. It is the author's hope that this book will serve both audiences by providing architecture developers with methods to address these integrated issues while providing operators a solid basis to develop skills to handle emergency situations with a proficiency that may one day save their lives. Chapters 2-4 deal more with issues of mission and spacecraft design, and as such, these chapters would be of most use to engineers and spaceflight architecture developers. Chapters 5-7 provide more of a background to the history and evolution of egress systems and procedures, material that would provide more a background role to the mission and vehicle designer but would be of paramount importance to the flight crew.

This book is intended as a reference for the professional student of architecture-level design of crewed spaceflight architecture. As such, it could support a stand-alone course or serve as a specialized topic of spaceflight systems engineering course at the upper-undergraduate or graduate level. It is also intended as a textbook to provide crew members and operators of spacecraft an understanding of the risks associated with pad, ascent, landing, and post-landing phases of space missions and a background of the methods that have evolved throughout the history of US, Russian, and commercial spacecraft to best mitigate these risks.

Table of Figures

List of Tables

ACRONYMNS AND ABBREVIATIONS

Acronym	Definition
AA	Access Arm
ACES	Advanced Crew Escape Suit
ADF	Australian Defense Force
AFB	Air Force Base
AIAA	American Institute Of Aeronautics And Astronauts
AM	Amplitude Modulation
AMSA	Australian Maritime Safety Authority
AMVER	Automated Mutual-Assistance Vessel Rescue System
ANT	Antenna
AOA	Abort Once Around
ASAP	As Soon As Possible
ASP	Astronaut Support Person
ASRK	Air-Sea Rescue Kit
ASTP	Apollo-Soyuz Test Project
ATCS	Active Thermal Control System
BAT	Battery
BCN	Beacon
BDA	Blast Danger Area
C-17	C-17 "Globemaster"
C-130	C-130 "Hercules"
CASARA	Civil Air Search and Rescue Association
CCAS	Cape Canaveral Air Station
CCDev	Commercial Crew Development
CCG	Canadian Coast Guard
CCGA	Canadian Coast Guard Auxiliary
CDR	Commander
CERA-40	Corrected ERA-40 Model
CEV	Crew Exploration Vehicle (a.k.a. Orion)
CFB	Canadian Forces Base
CIREN	Crash Injury Research Engineering Network
CM	Command Module
CMUS	Command Module Up-righting System
CO2	Carbon Dioxide
CSLAA	Commercial Space Launch Amendments Act of 2004
CSTS	Crew Space Transportation System
CW	Clockwise
DAEZ	Downrange Atlantic Exclusion Zone
DC	Direct Current
DCS	Data Communication System
	Decompression Sickness
DM	Descent Module
DoD	Department of Defense
EBA	Emergency Breathing Apparatus
EBS	Emergency Breathing System
ECLSS	Environmental Control And Life Support System

Acronym	Definition
ECMWF	European Centre For Medium-Range Weather Forecasting
EDL	Entry, Descent, and Landing
EDOMP	Extended Duration Orbiter Medical Project
EDS	Electrical Distribution System
ELS	Earth Landing System
ELSC	Earth Landing Sequence Controller
ELT	Emergency Locator Transmitter
EOS	Emergency Oxygen System
EPS	Electrical Power System
EVA	Extravehicular Activity
FAA/AST	Federal Aviation Administration Office of Commercial Space Transportation
FLT	Flight
FM	Frequency Modulation
FOM	Flight Operations Manual
GI	Gastrointestinal
GMO	Ground and Mission Operations
GMT	Greenwich Mean Time
GOES	Geostationary Operational Environmental Satellite
GSE	Ground Support Equipment
HF	High Frequency
HSIR	Human-Systems Integration Requirements
ICAO	International Civil Aviation Organization
IEEE	Institute Of Electrical And Electronic Engineers
IMO	International Maritime Organization
IRL	Indy Racing League
ISS	International Space Station
IV	Intravenous
JRCC	Joint Rescue Coordination Center
KSC	Kennedy Space Center
LAS	Launch Abort System
LDAR	Lightning Detection And Ranging
LEM	Lunar Excursion Module
LiOH	Lithium Hydroxide
LM	Lunar Module
LOC	Loss of Crew
LOM	Loss of Mission
LPU	Life Preserver Unit
LRS	Landing and Recovery Systems
MECO	Main Engine Cutoff
MET	Mission Elapsed Time
METAR	Meteorological Airfield Report
MIDDS	Meteorological Interactive Data Display System
MMH	Monomethyl Hydrazine
MR	Mercury Redstone
MRCC	Maritime Rescue Coordination Center
MRSC	Maritime Rescue Sub-Center
MS	Mission Specialist
MSC	Crewed Spacecraft Center
NASA	National Aeronautics And Space Administration
NASCAR	National Association for Stock Car Auto Racing

Acronym	Definition
NASS	National Automotive Sampling System
NCAR	National Center for Atmospheric Research
NCEP	National Center for Environmental Prediction
NEXRAD	Next Generation Weather Radar
NM	Nautical Miles
NOAA	National Oceanic And Atmospheric Administration
NPRV	Nitrogen Pressure Relief Valve
NSTS	National Space Transportation System
NTD	NASA Test Director
NTSB	National Transportation Safety Board
O2	Oxygen
OAA	Orbiter Access Arm
OM	Orbital Module
OMS	Orbital Maneuvering System
PEAP	Personal Egress Air Pack
PIC	Pilot In Command
PJ	Pararescue Jumper
PLS	Primary Landing Site
PLT	Pilot
PLS	Primary Landing Site
PLZ	Primary Landing Zone
PORT	Post-Landing Orion Recovery Test
PPRV	Positive Pressure Relief Valve
PPTS	Prospective Piloted Transport System
PS	Payload Specialist
PWR	Power
PYRO	Pyrotechnics
RAAF	Royal Australian Air Force
RAF	Royal Air Force (UK)
RAM-Z	Rigged Alternate Method - Zodiac
RCAF	Royal Canadian Air Force
RCC	Rescue Coordination Center
RCMP	Royal Canadian Mounted Police
RCN	Royal Canadian Navy
RCS	Reaction Control System
RHC	Rotational Hand Controller
RLY	Relay
ROCAT	Rocket / Catapult
RSLS	Redundant Set Launch Sequencer
RTAL	Retrograde Transatlantic Abort Landing
RTLS	Return To Launch Site
S/C	Spacecraft
SAR	Search And Rescue
SAROPS	Search and Rescue Optimal Planning System
SAS	Solar Array System
SCAPE	Self-Contained Atmospheric Pressure Ensemble
SEQ	Sequence
SM	Service Module
SMG	Spaceflight Meteorology Group
SOMS	Shuttle Orbiter Medical System

Acronym	Definition
SRB	Solid Rocket Booster
SRR	Search and Rescue Region
SSME	Space Shuttle Main Engine
STS	Space Transportation System
SWH	Significant Wave Height
SYS	System
TAL	Transatlantic Abort Landing
TBD	To Be Determined
UHF	Ultrahigh Frequency
US	United States
	Upper Stage
USSR	Union of Soviet Socialist Republics
VHF	Very High Frequency
VLV	Valve
WAM	WAve prediction Model

1 Introduction

In the early days of the crewed space programs of the United States and the Soviet Union, there were many unknowns. Sending a person into space was made dangerous not only by the hazards of the space environment itself, but also of the technology being developed to send the astronaut there. The rocket engines were not perfected and the propellants were just as apt to expend themselves in a large fireball at the pad as in the atmosphere en-route to orbit. While teams of engineers worked feverishly to make the rockets and spacecraft more robust and reliable, other engineers worked on protecting the flight crew should a problem arise. Every potential anomaly had to be considered but the greatest risks incurred on any space mission have always been on the launch and ascent phase and these risks are directly related to the failure modes of the launch vehicle. Launch abort systems have been designed in many different ways but they are the greatest mitigation method integrated to spacecraft to safeguard against critical launch and ascent failures.

Before the launch vehicle was even ignited, crews found themselves balanced atop millions of pounds of propellants, and sometimes surrounded in pure oxygen environments. If anything indicated any anomaly, the crew would need to get as far away from the rocket as possible as fast as possible. Uncertainties associated with launch vehicle reliability are largely mitigated through a launch abort system. Abort systems were integrated into space vehicles to largely mitigate the risks that the crew could perish should a launch vehicle failure occur during ascent. Launch abort systems have always been risky propositions; they are designed to pull or push the crew away from the launch vehicle with acceleration near the limit of human endurance. Failure detection is a paramount consideration as some launch vehicle failures gave a precious few fractions of a second to effect an abort. The requirements of NASA's now-cancelled Constellation Program called for a launch abort system that was 90% effective. In other words, it was acceptable to the program that the launch abort system would fail to save the crew in 10% of the attempts. Launch abort systems are systems artfully engineered to give the crew the best fighting chance in a desperate situation.

As the causes of catastrophic failures of the launch vehicle were not completely understood, engineers sought to detect failures with enough time before they could develop into a catastrophic environment in hopes of evacuating the crew. But pad-aborts (prior to launch) and ascent aborts (after launch) have historically relied on solid-fuel tractor rockets designed to pull the crew away from the doomed launch vehicle. These motors created an environment for the crew that was barely survivable; the launch escape system of the Orion spacecraft, integral to the Constellation Program and now being separately developed, was designed to impart 20 g's of force on the crew for 3 seconds. Getting caught within the exploding fireball of a launch vehicle was definitely not survivable, so launch abort systems were designed as a way to gain an added safety margin against launch vehicle anomalies. A successful abort gets the crew away from an immediately life-threatening situation, but places the crew in another challenging environment. For capsule spacecraft, the crew would be descending to Earth under parachutes, but the capsule could land upon any terrain along the ascent groundtrack. This terrain might include water, such as launches from Kennedy Space Center would find, or it could include land terrain such as would be encountered after an abort from a vehicle launching from Baikonur Cosmodrome in Kazakhstan. Since the crew has little ability to dictate the location of landing in such scenarios, there is landing risk. Furthermore, since a cost-effective contingency rescue capability cannot be implemented at all possible landing sites following an ascent abort, there will be delays in the arrival of the response team which would allow more time for a post-landing contingency to occur.

For capsule-based spacecraft, an egress prior to landing has been possible only through use of ejection seats, such as those that were used on the Soviet Vostok spacecraft, where the solo pilot would eject safely and as planned from the capsule upon descent. The American Gemini spacecraft also had powerful ejection seats for each of the two crewmembers for emergency egress upon ascent, though the system was never used in flight. Subsequent capsule-based spacecraft allowed for no such provisions, and egress operations needed to follow a safe landing. Winged space vehicles, such as the Space Shuttle, could only land on a suitable runway. The first four Space Shuttle missions (STS-1 through STS-4) had ejection seats available for the commander and pilot, though later Space Shuttle missions did not have these available and the only option available to Shuttle crews were to execute one of the predefined abort plans to landing at a suitable runway, either back at Kennedy Space Center or across the Atlantic Ocean at one of several Trans-Atlantic Landing (TAL) zones. If a suitable runway was unavailable, the only chance for the crew would be to conduct a risky "Mode VIII' egress operation involving a bail-out from a high-velocity while the vehicle is still in descent. The Mode VIII option was incorporated to the Space Shuttle orbiters following the explosion and loss of crew of the Space Shuttle Challenger in 1986.

A final scenario exists where the crew would be forced to abandon their vehicle. After the disaster of the Space Shuttle Columbia in 2003, which disintegrated upon re-entry following from damages incurred by the heat tiles that were caused by the impact of ice from the external tank upon ascent, NASA made provisions where a second Space Shuttle could be launched in time to rescue the crew from the doomed orbiter. The exchange of crew from the damaged orbiter to the undamaged one would require a complicated procedure involving numerous transferences of Extra-Vehicular Activity (EVA) spacesuits and equipment. For a further look into the procedures established by NASA for on-orbit emergency egress operations, the reader is referred to "Space Rescue, Ensuring the Safety of Manned Spaceflight" by David Shayler.

Emergency egress operations may also be triggered before launch through a rapid vehicle egress while the vehicle is still on the launch pad, as had been performed by the four Space Shuttle crews that faced Redundant Set Launch Sequencer (RSLS) aborts. The pad egress procedures that followed are detailed in Section 4.2.6. After the vehicle is launched, a pad-egress scenario is no longer possible. Emergency egress operations may then only be conducted upon descent or after landing. On some vehicles, egress operations may be conducted before landing during the descent phase of a mission, either upon re-entry or after a launch abort. For most vehicles, egress operations may only be conducted after landing. After landing, crews were vulnerable to anomalies that could occur before a rescue operation could be effected, and landings often did not occur where and when they were planned. If anything was to jeopardize the safety of the crew during or after landing, they would have to egress safely. Emergency egress operations are triggered by anomalous events that may occur upon descent (such as the decompression of the ill-fated Soyuz 11 or the introduction of toxic propellants into the Apollo capsule supporting the Apollo-Soyuz Test Project), upon landing, or at any time in the post-landing phase before the rescue operation has been completed.

Post-landing egress operations may follow nominal landings, where the spacecraft lands where planned, or they may follow a contingency operation where the crew has just performed a pad-abort, an ascent abort, or a rapid de-orbit operation. In such scenarios, the spacecraft may land upon terrain where it has not been designed to land and flight crews and rescue and recovery personnel must be trained to handle the situation effectively. For example the Soyuz spacecraft, though designed to land nominally upon the flat plains of Kazakhstan, has landed upon both mountainous terrain and in a body of water following in-flight contingencies. In both cases, the crew was successfully rescued. The success of these Soyuz missions resulted from a thorough planning of all 'what-if' scenarios prior to the missions. Decisions to commit to an emergency egress the spacecraft are rarely taken lightly as egress operations, especially in water, can involve more risk than might be incurred if the crew were to simply wait for rescue.

Emergency egress considerations are intimately tied with rescue and recovery architecture design and vehicle design. A more robust (and more costly) recovery architecture can guarantee recovery within a reduced timeframe and allow for less probability of post-landing contingencies. Ease of egress, from a vehicle standpoint, often comes at the expense of occupant protection during landing and overall vehicle size. Like everything in the engineering of interrelated systems, reasonable trades must be made, and data points have often come from previous crews that have faced dire post-landing scenarios at the time when the risks of the mission seem retired. A methodology of developing space architecture design and operational procedures for spacecraft rescue is detailed in Chapter 2. Vehicle design trades are presented in Chapter 3.4.

Egress systems and procedures have matured greatly since the early days of Vostok and Mercury. The first crewed spacecraft launched from the Soviet Union on April 2, 1961. The Russian Zvezda company had developed their own egress systems for the Vostok spacecraft that successfully carried the first six cosmonauts on orbital flights. The Vostok capsules used in these early Russian human space flights had ejection seats which were actually designed to be used nominally for the final stage of re-entry. The cosmonaut would re-enter the atmosphere in the capsule and descend under parachutes to an altitude of 6000 meters before ejecting from the capsule and descending under a personal parachute pack. The same ejection seat used for ejection from the spacecraft upon descent was also used for an ascent abort, should one be necessary and the spacecraft was at sufficiently low altitude and velocity for such an abort to be effective. In 1964, the Vostok spacecraft was renamed Voskhod and modified to accommodate multiple crew members, but with the sacrifice of an egress system.

Several thousand miles away the United States was preparing their first crewed space program. The Mercury Project launched six American astronauts, but unlike the Vostok spacecraft, the emergency egress of the solo occupant in a Mercury capsule was accomplished by the use of a parachute (with manually deployable reserve) which would lower the entire capsule to the earth. Aborts were handled by an escape rocket system which would separate the capsule from the rocket and lift it to an altitude where the parachutes could safely deploy. The impact was further attenuated by a large airbag like device which would deploy by separating the heat shield from the capsule and allowing its weight to pull down and inflate the bag. Initially conceived to land on land, Mercury was quickly adapted to water landings land in water. In this case, and for all American capsule-based spacecraft to follow, land nominal initial designs were traded for water landing vehicles to keep the vehicle within its mass allocation. After landing, the astronaut could egress through the side hatch, which was pyrotechnically actuated in all spacecraft except Alan Shepard's MR-3 spacecraft. Emergency personnel could also trigger the explosive hatch from outside the spacecraft by pulling on an external lanyard. However, emergency egress operations were generally planned through the antenna compartment, through which the pilot would pull himself after removing a small pressure bulkhead. This was a difficult and slow procedure but considered favorable in emergency scenarios.

In contrast to the Mercury Project, the Gemini Project that succeeded it used ejection seats for emergency egress. The seat was designed based on use during the pad, liftoff and late re-entry stages. Once in the upper atmosphere the seat firing handle, which was located between the astronauts' legs, was stowed so it wouldn't be inadvertently actuated in space. This seat used the most powerful ejection seat motor ever developed in the United States for propulsion, designed based on a projection of the radius of the expected fireball should the Titan II rocket explode on the pad. The projection required the seat to outrun the fireball for a distance of about 800 ft - fast enough that the nylon parachute would not be damaged by the heat pulse of the exploding fuel. The system was tested extensively by technicians using boilerplate capsule mock-ups which were launched at high speeds down sled tracks and seat ejections were filmed to verify high speed ejections were possible. Boilerplate capsules were also mounted in launch attitude on top of a high tower and seats were fired across the desert to prove the 800 foot mark could be reached, and the parachute system would deploy properly. After landing, the Gemini capsule proved to be stable in the water and it naturally oriented itself on its side. The parachute design configured itself before landing so that the capsule would splashdown on its side with the egress hatches for each of the two crew members facing upwards.

The Apollo spacecraft was developed from lessons learned during Project Mercury and Project Gemini to carry a crew of three astronauts to the moon and back, following the direction established by President Kennedy in 1964. Apollo reintroduced the launch abort tower that was used during the Mercury Program, though the Apollo variant was significantly more powerful. The tower was kept through first stage separation, after which the vehicle was traveling with enough velocity to render the system ineffective. An abort occurring before this time was termed a 'Mode I' abort, where the launch abort system could effectively pull the Apollo spacecraft away from the Saturn launch vehicle. After first stage separation, the launch abort system would be ejected from the spacecraft and any subsequent launch aborts would be performed through use of the Service Module (SM), which could push the spacecraft away from the launch vehicle. This type of abort was termed a 'Mode II' abort.

Upon water landing, the Apollo spacecraft had the misfortune to be stable in two different orientations; it could land right side up (a 'Stable One' orientation) or it could land upside down (a 'Stable Two' orientation). Obviously the latter was not desirable since it produces great stress to the crew and complicates egress and rescue operations. To account for this possibility, Apollo capsules were fitted with an uprighting bag system, called the Command Module Uprighting System (CMUS) which would inflate in the event of a 'Stable Two' landing and force the capsule into a Stable One orientation. Almost half of all Apollo missions landed in a Stable Two. Once in a Stable One, the crew could egress through either the side hatch or the docking hatch, located atop the spacecraft. In the event the up-righting system failed, the crew would have been forced to partially flood the spacecraft and egress through the docking hatch, with a very real risk of being pulled down by a sinking spacecraft.

Meanwhile in the Soviet Union, the six successful flights of Vostok and two of its multi-crewed successor, Voskhod, led to the development of the Soyuz spacecraft, which became the staple of the Russian Space Program. There have been three major contingencies during the descent and landing phase of Soyuz Missions. A parachute failure and a premature opening of the post-landing ventilation system claimed the lives of Soyuz 1 and Soyuz 11, respectively. Later, an errant reentry trajectory caused the Soyuz 23 spacecraft to land in the freezing waters of Lake Tengiz. Despite these failures, the Soyuz spacecraft grew to become world's safest, most cost-effective human spaceflight system.

Emergency recovery of the crew of the Soyuz can be performed from the launch pad or during ascent through use of a launch escape system. These emergency systems have been used twice. During the Soyuz T-10-1 mission in September 1983, a pad abort was successfully executed two seconds prior to a catastrophic explosion of the launch vehicle. Then, in April 1975, a high-altitude ascent abort was executed which resulted in only minor injury to the sole cosmonaut of the Soyuz 18-1 flight. To handle the possibility that an on-orbit contingency could force the crew to prematurely de-orbit, the Soyuz Program had identified three emergency landing zones that were considered to have a predictable enough terrain to have a high probability of a safe terrestrial landing: 1) the Sea of Okhotsk, 2) the steppes of Kazakhstan, and 3) the plains of the United States. In the event of an emergency landing on foreign soil, the Descent Module (DM) is labeled with instruction in both Russian and English instructing a potential rescuer to 'Take Key – Open Hatch' with further markings warning of the pyrotechnically activated hatch: "Attention! Step Aside! The cover may be jettisoned!" and just under the USSR/CCCP markings: "People Inside! Help!"

The US Space Shuttle, and the Soviet 'Buran' that followed and borrowed many of the design elements of the Space Shuttle, abandoned the capsule designs of preceding spacecraft in favor of a winged design. The immediate benefit was a huge increase in the available cross-range capability of the spacecraft; the orbiter could target and land on one of a number of suitable landing strips instead of being mercy to the ballistic trajectory that capsule designs must follow. The disadvantage of the Shuttle's winged design was that, if a landing strip could not be reached, the high approach speeds meant landing elsewhere was almost guaranteed to not be survivable. After the Challenger accident of 1986, the Mode VIII 'bail-out' capability was integrated to each of the orbiters, though many engineers felt pessimistic about the ability of Shuttle crews to safely egress in such a manner.

In 2004, the Constellation Program was conceptualized by NASA as a grandiose four-phase space architecture that would maintain the crewed presence of the United States in space by first continuing to support the International Space Station (ISS) and later by establishing a permanent lunar colony at the South Pole of the Moon, and eventually leading a crewed mission to Mars. Constellation was also to rely as much as possible on heritage space systems, and the conical design of the Orion spacecraft derived itself from the design of Apollo. The larger Orion spacecraft, however, was initially designed to carry a crew of six to the ISS, though overestimates of the capability of the Ares launch vehicle, combined with the abandonment of the Mars Design Reference Mission, eventually forced the Orion spacecraft to downsize to accommodate a crew of four. Like Apollo, Orion relied on a tractor emergency escape system that could pull the spacecraft from the launch vehicle for Mode I aborts or, use the SM motor to push Orion away for Mode II aborts. These aborts could even be targeted so that the ensuing splashdown would not occur in environments where recovery operations could be most hazardous, such as the North Atlantic. These targeted aborts were termed 'Mode III' aborts. Also like Apollo, Orion had two stable orientations in water and thus relied on the use of an up-righting system to avert a post-landing scenario where the crew would be suspended upside down in an inverted spacecraft. Egress could be conducted, either assisted by a rescue team or unassisted, through either the side hatch or the docking hatch atop the vehicle. Unless the crew needed to self-egress in the event of a post-landing emergency, recovery teams would stabilize the capsule through use of a flotation collar and a sea anchor before extracting the crew.

By the conclusion of 2010, the Constellation Program had been cancelled and NASA took a direction in favor of supporting cutting edge and 'game-changing' technologies while entrusting commercial enterprise to develop the spacecraft and launch vehicles to send people to space. These next-generation spacecraft derive from various designs and mission requirements, from glider-based suborbital spacecraft as are being developed by Virgin Galactic and XCOR to orbital spacecraft of capsule design, such as the Dragon spacecraft being produced by SpaceX, and orbital spacecraft relying on lifting bodies, such as the Dream Chaser spacecraft being produced by Sierra Nevada Corporation. Various companies had been awarded government funding through NASA's Commercial Crew Development Program (CCDev), with SpaceX and Boeing eventually receiving contracts to supply the International Space Station.

At time of this writing, nominal recovery plans are in a state of relative maturity though contingency rescue plans, historically relying upon government run institutions, such as the Department of Defense (DoD), and international agreements with foreign countries possessing Search and Rescue (SAR) capability. It is uncertain how contingency rescue plans will be integrated into the mission planning of these new commercial spacecraft. A projection of post-landing and egress operations concerning commercial space vehicles and crew is provided in Chapter 8.

2 Post-Landing Operational Considerations in Mission Planning

The most important factor in determining the risks associated with the post-landing phase of any crewed space mission is the environment into which the spacecraft lands. For nominal missions, this is rarely a factor since the primary landing zone will be of environmental conditions that the rescue and recovery methodology has been well validated against. Atmospheric conditions, and sea conditions if the spacecraft is to land in water, rarely change significantly over the period of time from the de-orbit burn to landing, so once the decision is made to re-enter, the spacecraft should experience no environmental 'surprises' and the rescue and recovery team is already 'go' to proceed with the rescue operation. Historically, nominal recovery operations take less than two hours to conduct, but two hours is plenty of time for a post-landing contingency to arise that would force the crew to conduct an unassisted emergency egress. If the landing will be in water, such as the Mercury, Gemini, Apollo, and Orion spacecraft were designed to do, the sea conditions would be below the threshold limit that crews have demonstrated the ability to conduct a safe egress. A water-landing spacecraft would not intentionally target seas of conditions where both rescue operations and emergency egress procedures have not been well validated.

Contingencies, however, may cause the spacecraft to land in environments that are not ideal and these environments may be in excess of what rescue teams can conduct safe rescue operations in or in excess of what the crew may conduct safe egress operations in. This is especially true for water landings where the sea motion can complicate the recovery operation and sinking becomes a credible risk. Proper mission planning will minimize the possibility that the spacecraft will land in an environment of excessive conditions, and if it does, the most effective procedures will have been established to protect the crew. However, trades are often made between the operability of the spacecraft architecture and risk as the increasing of launch opportunities in a given timeframe often means accepting more risk. And although the making of a more robust rescue and recovery may reduce post-landing risk to a degree, such investments often come at great costs and a reasonable compromise must be made.

There are four main phases of flight that have or could potentially involve assisted or unassisted crew egress operations: 1) pre-launch, 2) post-landing, 3) descent, and 4) on-orbit operations. Pre-launch emergency egress operations are often triggered by anomalies in the integrity or stability of the launch vehicle and require the crew to either egress from the pad via an escape system, such as were performed by the five Space Shuttle crews that experienced an RSLS abort. Post-landing emergency egress operations are often triggered by the introduction of a hazard to the crew in the post-landing phase of flight. The post-landing phase may be part of a nominal end-of-mission landing, a contingency end-of-mission landing, or after a pad or ascent abort. The post-landing phase may also be on water or land. A major concern of post-landing contingencies after an otherwise nominal mission would be the state of deconditioning of the crew. A major concern of post-landing contingencies following a pad or ascent abort would be the greater stresses imposed on the vehicle, the heightened probability of injury to a crew member, and the greater probability of landing in a less predictable and potentially more hazardous environment. There have been several exceptions where egress is a consideration during the descent phase following a launch abort or where landing the vehicle would be deemed unsafe (e.g. Space Shuttle 'Mode VIII' abort), or where the vehicle was simply not designed to land in a way to protect the occupant, such as were performed on the six Vostok missions. Several vehicles also had functioning ejection seats where a rapid egress could be performed at low altitude and low velocity in the ascent or descent phase of flight, such as were on the Gemini capsules and the first four Space Shuttle missions. Finally, emergency egress operations performed while on-orbit require coordination with a second vehicle (e.g. rescue vehicle, Space Station) and are rarely time critical.

This section describes the methodologies used to predict what sort of environments the crew may face in the post-landing phase of a mission based on the sum of the failures that could happen from launch to landing. By understanding the various environments that the spacecraft could land in, and the associated probabilities of landing within each environment, an effective rescue and recovery system and an effective post-landing contingency plan may be developed. The rescue and recovery assets detailed in this section are those that were used by NASA for the support of the Space Shuttle and were the planned means of rescue and recovery operations for the Constellation Program. The Space Shuttle, as a winged spacecraft, has much more ability to target specific landing zones following an abort and is therefore not as dependent on a global rescue and recovery architecture as a capsule-based spacecraft would be. Capsule-based spacecraft largely follow ballistic trajectories following ascent aborts and have little cross-range capability once committed to de-orbit, so rescue and recovery plans must consider that there could be a problem at any time that could place the spacecraft in almost any point on Earth, though we will see that there is a much greater probability that the spacecraft could abort into some regions, such as the ascent groundtrack, than others.

2.1 Nominal Rescue Operations

Before the Space Shuttle program, all American space vehicles were nominally designed to land in water. All Russian and Chinese space vehicles have been designed to land nominally on land. Since 1981, all Space Shuttle missions have landed at one of the three end-of-mission landing sites: Kennedy Space Center, Edwards AFB, and White Sands, NM, with the exception of the ill-fated final missions of Challenger and Columbia. All Russian vehicles have landed on land in what is now the nation of Kazakhstan (though several landed in off-nominal ways after experiencing serious contingencies) with the exception of Soyuz 23, which errantly landed in the frozen Lake Tengiz.

2.1.1 Planning and Preparing of Primary Landing Sites

The selection of landing sites considers many factors including the orbital ground track, the expected downrange distance from the de-orbit burn to landing, the terrain and environments about the landing site, the presence of populated areas that might be at risk of falling debris, the maintenance of communications links, availability of recovery assets, and any political constraints. Because of the ability of winged spacecraft to alter their course to target a specific landing site, only the environments at and about the landing site are considered. The landing locations of capsule-based spacecraft, however, are not as confined as winged vehicles and are thus generally selected within a region of relatively benign and constant terrain so that de-orbit or re-entry errors would have only a small impact in the overall landing and post-landing environment.

For the first two suborbital flights of Project Mercury, the splashdown locations were fairly well constrained by the downrange capability of the Redstone launch vehicle. Once the Atlas booster provided orbital flight to the subsequent four Mercury missions, the splashdown location was moved to locations farther east. Re-entry was a little known science and a major risk driver was the possibility that an errant reentry operation could cause the spacecraft to hit land, which could cause serious injury or death to the astronaut. The second orbital Mercury mission, MA-7, was commanded by Commander Scott Carpenter. Carpenter overshot his reentry and splashed down 250 miles (400 kilometers) from his target, near the Virgin Islands. Largely due to this event, NASA decided to incur the additional logistical expenses of landing the remaining two Mercury spacecraft in the Pacific Ocean. Confidence of the ability to target landing areas was quickly restored by the start of Project Gemini, so landing areas were once again targeted in the Atlantic Ocean. Once Project Apollo spacecraft started to leave Earth orbit, the added error associated from the higher velocities and the uncertainties associated with the earth transfer orbit made the larger and calmer Pacific Ocean desirable. Further, the 'skip' re-entry technique, an unproven method of bleeding off the excessive speed of a lunar return, was considered but it was recognized that such a return could cause huge errors in the descent trajectory and landing location if the initial entry into the Earth's atmosphere was only a little off. For these reasons, splashdown locations were again selected in the Pacific Ocean. After Apollo, the Space Shuttle permitted very precise control over landing operations so mobile rescue capability was only needed for contingencies. Soyuz missions, however, have all targeted the Kazakh Plains and re-entry errors, such as ballistic re-entries triggered by attitude control system problems, have necessitated the use of flexible and mobile rescue and recovery assets.

Preparations for a spacecraft landing typically begin weeks or months prior to the actual event. These preparations include equipment preparations and mission briefings, but also Notices to Mariners and Notices to Airmen must be submitted several days prior to the landing in order to alert sea vessels and aircraft of water and airspace restrictions that will be needed to be established about the landing site. For capsule designs, the actual splashdown site will vary and be dependent upon such factors as ground track, weather, and other considerations. This means that the actual location of splashdown or landing is frequently not determined until the mission is underway, though the general location may be well established.

2.1.2   Water-Nominal Spacecraft

American spacecraft until the Space Shuttle era have always landed in water. The decision of NASA to land in water was likely driven by 1) a desire to use lighter spacecraft or maintain more mass margin on the spacecraft, and 2) the availability of relatively calm waters at the lower latitudes of the Florida peninsula. Mercury, Gemini, and Apollo spacecraft were all designed to have water nominal landings. The greatest risk of one of these capsules hitting land would have occurred following a pad abort into easterly winds that could blow the capsule back onto the Florida coast. The Orion capsule was initially designed to have land-nominal landings but mass constraints imposed by the Ares 1 launch vehicle forced the design to be water-nominal.

As all capsule-based NASA spacecraft have all landed in the open ocean, a recovery ship is an essential part of the mission architecture. Historically, this ship has been provided though agreement with the DoD, though the Constellation Program had been investigating the use of commercial ship acquisition for recovery operations. The SpaceX Dragon capsule relies exclusively upon commercial ships for nominal recovery operations. The only Russian spacecraft to land in water was Soyuz 23, resulting from an off-target landing.

2.1.3   Preparations for Landing

Landing support for capsule based spacecraft have been generally similar from the perspective of the recovery ship, though smaller capsules relied upon hoisting via helicopter while larger capsules relied upon stabilization and hoisting by shipboard crane.   Prior to splashdown, the primary recovery ship would position itself just outside of the landing zone.   Here, rescue team personnel inventory, inspect, and calibrate support equipment required for the landing operation on the recovery ship.   Once these actions are complete, the NASA team loads equipment and personnel onto the ship, which include specialized mission expertise such as swimmers, vehicle handlers, medical personnel, science support personnel, safety and public affairs personnel, a weather officer, and communications representatives.   The Recovery Team Leader then provides a NASA recovery mission briefing that provides the timeline for departing port, the sequence of events for the entire at-sea portion of the recovery effort, weather forecast, ship Captain's comments, safety considerations, and other topics at the discretion of the Recovery Team Leader. If the ship has a suitable landing deck for helicopters and the helicopter aircrews are qualified, current, and approved to land on the deck, then a separate pre-sail air operations brief also takes place.

Mercury Nominal Landing

The Mercury Capsule was designed to activate its landing system once its redundant pair of onboard pressure sensors detected an ambient atmospheric pressure of 21000ft (6.4km), at what point the drogue parachute would be deployed to stabilize the capsule and decrease the velocity in preparation of deployment of the 19.2m diameter main parachute, which were triggered at an altitude of 10000ft (3.05km) after the drogue chute and antenna package were jettisoned.

The main parachute would be deployed initially in a reefed condition that provided drag from a cross-sectional area of only 12% the total deployed area of the parachute. Four seconds later, the parachute would extend to its full diameter, thus minimizing the shock experienced by the parachute and the vehicle. Then twelve seconds following the deployment of the main parachute, a landing bag was deployed. The purpose of the landing bag was to both diminish the impact force to a manageable 15g and then to act as a stabilization anchor while the capsule was afloat, facilitating rescue and recovery operations.

Gemini Nominal Landing

An initial goal of the Gemini Project was to develop the capability to land upon land, thus saving operational costs by allowing for refurbishment of the vehicles. Of the investigated concepts, the furthest developed included one employing a paraglider with landing gear, and another using a parasail combined with landing rockets. Eventually, this design objective was scrapped and the Gemini capsule would land in the water as the Mercury capsule had before it.

Similar to that of the Mercury capsule, the Gemini landing system deployed its parachute system sequentially, though Gemini used a series of three parachutes with the main parachute deployed in a manner to land the spacecraft on its side. The Gemini spacecraft deployed a 3.3m high-altitude drogue parachute first at an altitude of 50,000 ft (15.2km), followed by a 6.4m pilot parachute at an altitude of 10,600 ft (3.2km), and a 25.6m main parachute at an altitude of 9800 ft (2.9km), each parachute sequentially deployed at altitudes determined by an onboard pressure sensor system. Unlike the Mercury capsule, the main parachute was a ring sail-type parachute that would drop to a two point configuration that would pitch the spacecraft at a 35° angle to the horizontal that provided greater safety to the astronauts as they impacted the ocean. The angle at which the Gemini capsule hung allowed the spacecraft to impact the water at the edge of its heat shield, reducing landing loads through a reduction of the initial contact area between the capsule and the surface. The Gemini descent and landing is depicted below in Figure 2.1.

After landing, the parachutes were severed from the spacecraft and recovery aids were deployed. A failure of the drogue parachute could be overridden as the crew could manually activate an alternative deployment sequence that would free the failed drogue and deploy the pilot parachute at an altitude of 3.2km, from which a nominal landing could still be achieved.

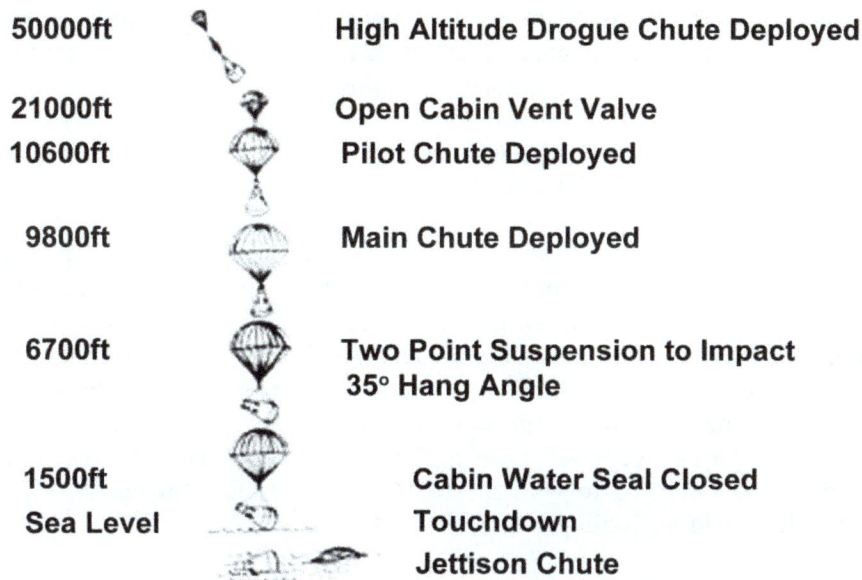

50000ft		**High Altitude Drogue Chute Deployed**
21000ft		**Open Cabin Vent Valve**
10600ft		**Pilot Chute Deployed**
9800ft		**Main Chute Deployed**
6700ft		**Two Point Suspension to Impact** **35° Hang Angle**
1500ft		**Cabin Water Seal Closed**
Sea Level		**Touchdown**
		Jettison Chute

Figure 2.1 Gemini Descent and Landing Sequence

Apollo Nominal Landing

The Apollo spacecraft landed in water much as its Mercury and Gemini predecessors, but the size and mass of the spacecraft led to a larger, three-parachute design. The Apollo Earth Landing System (ELS) consisted of the parachute subsystem, two Earth Landing Sequence Controllers (ELSCs), and the apex cover jettison mechanism, and was designed to be one fault tolerant in all critical systems, including the parachute system. If one of the three parachutes were to fail to deploy, the landing velocity would be marginally increased to 10.2 m/s from a nominal 8.5 m/s, well within safe limits of operation. Once in the water, the spacecraft had two modes of stability; the Apollo Command Module (CM) could orient itself in one of two stable orientations: a right-side-up ('Stable One') orientation or an inverted ('Stable Two') orientation. To address the possibility of a Stable Two landing, an uprighting bag system was developed so that the spacecraft could right itself.

The parachute subsystem is comprised of two fist-ribbon-type nylon drogue parachutes, 13.7 feet (4.2 m) in diameter; three ring-slot-type nylon pilot parachutes, 7.2 feet (2.2 m) in diameter; and three ring-sail-type nylon main parachutes, 83.5 feet (25.5 m) in diameter. After the logic circuitry is armed as the vehicle descends past 24000ft, the ELSC automatically senses altitude and initiates deployment of the parachutes at the proper time by using both barometric switches and time-delay relays. The drogue parachutes are then deployed following a 14 second time delay into a reefed condition for 8 seconds before a pyrotechnically-activated cutter allows the parachutes to fully open. In this configuration, the CM will descend to approximately 10,000 feet where a barometric switch will release the drogue parachute prior to the mortar deployment of the three main parachutes. The main parachutes are reefed for 8 seconds before the parachutes fully inflate to lower the CM safely to landing, after which the main parachutes are released from the CM through use of reefing line cutters, three of which are employed on each of the two reefing lines for the main parachutes. The events of the Apollo descent are listed with their triggers and associated descent velocities in Table 2.1.

In the event of an abort using the Launch Escape System (LES), the descent and landing sequence assumes the crew may be unresponsive and would proceed automatically. If the launch escape tower is still attached, it is jettisoned with the apex cover at 24,000 feet, and the landing sequence proceeds in an otherwise nominal manner. Backup emergency switches are provided for manual control of the apex cover jettison and parachute deployment operations. The backup circuitry provides more than redundancy of a critical system; it also can be used to delay deployment of the parachutes if there is possibility that a nominal deployment could permit the CM could blow back and impact land following a pad abort.

After landing, the CM uprighting system is manually activated only if the CM is in a Stable Two orientation. This system consists of three inflatable air bags, two relays, three solenoid control valves, two air compressors, control switches, and air lines. The uprighting bag system is activated manually through three switches that independently inflate each of the spacecraft's three 24 ft^3 air bags. Once the CM is uprighted, the three activation switches can be set to a mode that isolates and maintains the pressure within each uprighting bag. Post-landing ventilation for the crew is supplied through two vent valves in the forward access hatch cover. These vent fans could be turned on after landing to vent the CM to the outside atmosphere, eliminating CO_2 buildup as well as to providing cooling.

Event	Trigger	Descent Velocity
Apex Cover Jettison	24,000ft barometric switch (24,900 to 21,500 feet)	425 ft/s (129 m/s)
Drogue Parachute Deployment	Two second time delay from 24,000 ft barometric switch close	410 ft/sec (125 m/s)
Drogue Parachute release and Main Parachute Deploy	10,000 ft barometric switch close (10,950 to 9,100 ft)	225 ft/s (69 m/s)
Reefed Opening of Main Parachute System	Eight Second Time Delay after line stretch	235 ft/s (72 m/s)
Disreef of Main Parachute System	8,400 feet (+/- 500 ft)	110 ft/s (34 m/s)
Touchdown (3 main parachutes deployed)	Roughly 5 minutes after main parachute deployment	28 ft/s (8.5 m/s)
Touchdown (2 main parachutes deployed)	Roughly 4.2 minutes after main parachute deployment	33.5 ft/s (10.2 m/s)

Table 2.1 Apollo ELS Events and Descent Rates

POST-LANDING OPERATIONAL CONSIDERATIONS IN MISSION PLANNING

To assist in the rescue team, post-landing recovery aids can be deployed by the crew. These aids consisted of a sea dye marker, swimmers umbilical, High Frequency (HF) recovery antenna, and a flashing beacon light. The sea dye marker and swimmers umbilical are deployed automatically when the recovery antenna is deployed and tethered to the CM forward compartment deck. The sea dye marker would last approximately 12 hours. For long-term post-landing operations, a raft and survival kit are available to the crew.

Orion Nominal Landing

The Orion capsule was designed to land in a similar way to Apollo. Two 23ft (7m) diameter conical ribbon drogue parachutes are mortar deployed at 45000ft altitude to provide initial deceleration and stabilization, followed by two 32ft (9.75m) ringsail parachutes. Once the vehicle descends to 8000ft AGL, Orion's three 118ft (36m) diameter main parachutes are extracted and deployed as the forward bay cover is jettisoned. The main parachutes are attached to the CM via a 28° hang angle and provide a final descent to touchdown. If all three main parachutes are functional, the touchdown velocity would be 24 ft/s (7.3 m/s), a significant reduction from the 28 ft/s touchdown velocity of the Apollo CM. If one parachute on the Orion system had malfunctioned, the touchdown velocity would be 33 ft/s (10 m/s), still within the human factors limits (see Section 0) when considering the impact protection of the vehicles crew couches and impact absorption system.

Orion Rescue Operations

The operational plan of a nominal rescue of the Orion capsule is detained here. The plan is not very dissimilar from previous planning to rescue the crew from water-landing spacecraft, so it can be considered the most evolved concept of operations developed for a water-landing vehicle of a national space program, but the concepts build on the heritage established by the Mercury, Gemini, and Apollo projects.

The Constellation Program, having confidence that a splashdown location could be targeted with only 10km error, selected a location off the coast of Southern California for nominal splashdowns. This area was easily and readily accessible by assets based at the North Island Naval Air Station, so the DoD would be able to provide sufficient helicopters to support medical evacuation of all astronauts. DoD also would provide C-130 aircraft for in-flight refueling of the helicopters and to act as an airborne SAR coordinator. For the return segment of Constellation lunar missions, DoD adds a long range airborne alert aircraft (typically a C-17) for immediate contingency response, which would depart directly from the aircraft home station on the day of the planned splashdown. Training of the entire recovery team, including the ship's crew, is essential so that the recovery team would be proficient in all required tasks prior to splashdown.

Once the landing zone location and landing time are established, the recovery ship is deployed and the Recovery Team Leader directs initial on-scene actions, ensuring the landing zone is clear of other vessels. At this time, the DoD helicopters and fixed-wing aircraft that would be used in the event of a medical evacuation fly to a designated station keeping point in the near vicinity of the landing zone. With both DoD and NASA recovery assets in-place, the Recovery Team Leader announces that the combined recovery team is ready to support spacecraft splashdown.

With the landing zone and airspace clear, recovery ship in-place, and flight rules met (e.g. systems, weather, etc. are "go" for landing), the astronauts accomplish a nominal de-orbit burn to begin the landing sequence. An exclusion zone is established around the landing site with a 10 km radius to provide an unobstructed landing zone for both the SM and the CM, and to ensure NASA team safety. Once the splashdown location is refined as the spacecraft descends under the main parachutes, the Recovery Team Leader permits recovery vehicles to enter the exclusion zone when it is safe to do so. At this point, splashdown is imminent and all recovery vehicles proceed toward the updated splashdown location to begin recovery operations.

After splashdown, the astronauts establish voice communication with recovery forces, perform 'auto-safing', and remain suited and seated in the spacecraft until recovery personnel open the hatch. The 'auto-safing' feature provides a safe environment for the astronauts and recovery personnel and takes approximately 15 minutes to complete. Once the safing is complete, avionics are shut down and the crew switches to a combination of S-band emergency communications, though a choice of line-of-sight only VHF, UHF, or FM radios is available for communication with on scene recovery forces.

With splashdown confirmed, surface recovery forces converge on the CM from an upwind direction, avoiding possible exposure to toxic gasses, and standoff at a safe distance until vehicle safing is confirmed complete. Simultaneously, rescue helicopters and fixed-wing aircraft enter the immediate recovery area to evaluate the situation and locate and mark the spacecraft, parachutes and forward bay cover. Overhead, a videographer onboard a C-130 documents the condition of the spacecraft and the recovery process until the spacecraft and crew are safely aboard the recovery ship.

With a safe vehicle confirmed by telemetry and/or voice communication with the astronauts, the Recovery Team Leader directs the first responders to "sniff" the atmosphere to determine if any evidence of hazardous chemicals is present and to visually inspect the spacecraft for any signs of leaks or damage. If chemical hazards are detected, the Recovery Team Leader takes appropriate action and notifies the crew. Otherwise, the first responders attach a sea anchor to slow the capsule drift rate, and then wash the thrusters with sea water sprayers to eliminate any residual hazards. The stabilization collar is then attached which helps attenuate the motion of the spacecraft in the water and provides first responders a platform to access the spacecraft side hatch, windows, forward bay and if required the docking hatch.

Meanwhile, the recovery ship is maneuvered into position to prepare to lift the capsule. The lifting process is accomplished via well-deck, elevator-deck, A-frame, launch and recovery system, crane operations or other suitable human rated lifting systems. Regardless of the lifting method, the lifting hardware, application software, and associated processes are suitable to lift the capsule with the astronauts remaining in the vehicle. The crew has the option to run the capsule snorkel fan throughout the lifting operation.

After the capsule is safely secured on the deck, recovery forces position themselves for a final "sniff" check prior to accessing the capsule side hatch. Once the area is deemed safe, the access stand is locked in-place and recovery team members open the side hatch. The lead surgeon climbs the access stand and conducts the initial medical overview of the astronaut's health and well being. Once the initial medical overview is complete, other NASA team members access the side hatch and begin assisting each astronaut out of the vehicle utilizing handholds or lifting points on the crewmember's suits. Each astronaut is escorted out of the CM in a supine position, down the access stand, and then to a post-flight processing area. NASA team members then assist suit doffing and then help the astronauts to a medical examination room where they undergo medical evaluations and required science procedures while remaining under the care of medical personnel until released by the lead surgeon.

If the astronauts require medical attention that exceeds the capability on the ship facility, the lead surgeon may request medical evacuation support from DoD. The preferred method to load the patient(s) is by landing on the ship but if that is not possible, the patient(s) is hoisted to the helicopter while the helicopter is in a low hover. The typical medical manifest on the helicopter is one flight surgeon and two Pararescue Jumpers (PJs) or Emergency Medical Technicians to care for two astronaut patients. With the patient(s) loaded, the aircrew departs the recovery ship en-route to the intermediate / definitive medical care facility determined by the lead surgeon on the recovery ship.

2.1.4   Land-Nominal Spacecraft

Soviet and Russian spacecraft have always intended to land on land, and all but one did. Vostok, Voskhod, and Soyuz spacecraft have all been designed to be land nominal. The desire of the Soviet Union to design a spacecraft that would land safely on land probably came from 1) the high latitude of the country where domestic launches would be placed in a moderately high inclination, 2) the availability of large flatlands in the Kazakh SSR, and 3) the greater relative secrecy of the political climate that would drive an interest of having both launches and landings within national borders.

All American astronauts that have flown on a domestic vehicle after 1980 have flown on the Space Shuttle. The winged design of the Shuttle has allowed for all missions to be planned around a terrestrial landing on a suitable landing strip, though in the event of a contingency, there were over 60 landing zones upon which the Shuttle could land on. Current designs of the Sierra Nevada Dream Chaser as well as future variations of the SpaceX Dragon capsule will be designed to land upon land, so it is likely that the trend of spacecraft in the future will be towards terrestrial landings, largely because of the logistical cost savings, greater degree of reusability, and less involved contingency planning.

Vostok Nominal Landing

The "Vostok" ('East') spacecraft lacked a landing system. Obviously, it was essential that the cosmonaut be able to egress the spacecraft in time lest he perish as Vostok impacted the ground. At an altitude of seven kilometers, the main hatch of the capsule was jettisoned and seconds later, the pilot ejected. The unoccupied Vostok DM would then descend to an altitude of 4,000m, where a drogue parachute would be deployed and then a main parachute would follow at an altitude of 2,500m. Meanwhile, the cosmonaut would separate from the ejection seat at 4,000m to descend via a personal parachute system. In the event of a failure of the main parachute, a reserve parachute could be deployed. There were times when the crew-member experienced g-forces in excess of 10g's. According to Yuri Gagarin:

> "There was a moment, about 2-3 seconds, when data on the control gauges started looking blurry. It was starting turning gray in my eyes. I braced and composed myself. It helped, everything kind of returned to its place."

After landing, the cosmonaut would await the rescue team and initial medical tests would be performed before the cosmonaut was airlifted back to Moscow for extended examinations and de-briefings. If an emergency had occurred, a special team of doctors were trained to parachute into the landing zone to assist.

Vostok flew six missions before being upgraded to a variant called "Voskhod", or 'Sunrise, in Russian. The Voskhod spacecraft were basically Vostok spacecraft that could accommodate two cosmonauts with a backup, solid-fuel retro-rocket added onto the top of the DM. There was no provision for escape for the crewmen in the event of a launch or landing emergency. Only two Voskhod missions were to fly before the advent of the Soyuz Program, which was to start as tragically as the Apollo Project.

Soyuz Nominal Landing

The Soyuz landing sequence begins with the deployment of the pilot and drogue chutes at an altitude of about 10km. Once the DM descends to about 5km, the main parachute is deployed. If the main parachute were to fail, a reserve parachute may be deployed which would entail the separation of the main parachute. Under the 570 square meter reserve parachute, the descent rate of the DM is 10 meters per second, as opposed to the nominal descent rate of 8 meters per second of the main parachute. The descent profile for the Soyuz DM is illustrated in Figure 2.2.

Once the DM is at an altitude of 3km, the heat shield is ejected. The landing is assisted by the firing of four solid-rocket retro rockets that are activated when a gamma-ray altimeter detects the spacecraft to be at an altitude of 1.5m above the ground. The retro-rockets slow the final descent rate to approximately 2.5 meters per second, a rate easily managed by the cushioned seats. If the retro-rocket system were to fail under the reserve parachute, the landing would be at a velocity of 10 meters per second. The probability for an injury that the crew-members might receive under such a rough landing would be very dependant upon the nature of the terrain the DM lands in. After landing, the main parachute is jettisoned and hatch may be opened by the crew or jettisoned. In consideration of the possibility of injured crew-members, instructions in both Russian and English are placed to direct a rescuer to open the hatch.

The main landing zone for Soyuz missions is in Kazakhstan, between Tselinograd and Dzhezkazgan, as shown in Figure 2.3. Rescue and recovery of the crew and the spacecraft is coordinated by the Federal Aerospace Search and Rescue Administration. Landing is often times so that both the de-orbit burn and landing are both in daylight so that the crew can adjust the attitude of the spacecraft just prior to touchdown.

The recovery operation relies upon a number of aircraft and all-terrain vehicles. Before landing, aircraft are used to spot the DM and then relay its position via radio to the ground vehicles. Communications are maintained with the crew on a separate frequency. After landing, recovery personnel approach the DM to render assistance in opening the hatch and assessing the medical condition of the crew. Then, the DM is up-righted so that the crew may exit via a slide that helps place each crew-member in a seat to be moved to a tent from where more detailed medical evaluations may be performed. As crew-members may be injured from landing or highly deconditioned from extended stays in orbit, they are given assistance through every step of the egress procedure. Finally, the crew is flown back to Baikonur and then to Moscow and the DM is secured to a ground vehicle, fuel and pyrotechnics removed, and transported to Energiya at Korolyov, where a complete assessment of the vehicle may be conducted.

Future Soyuz spacecraft may be equipped with an improved automated landing system that would allow for parachute deployment at lower altitudes and increase the overall landing accuracy. The addition of two more retro-rockets could reduce the final descent rate to only one meter per second. Russian officials hope that such modifications would allow for landing within Russia, instead of Kazakhstan. Currently, the Chinese have designed their Shenzhou spacecraft relying heavily on Soyuz design elements, though improvements to the landing system, including a larger main parachute and stronger retro-rockets, have already been implemented to reduce the landing impact to a mere one meter per second.

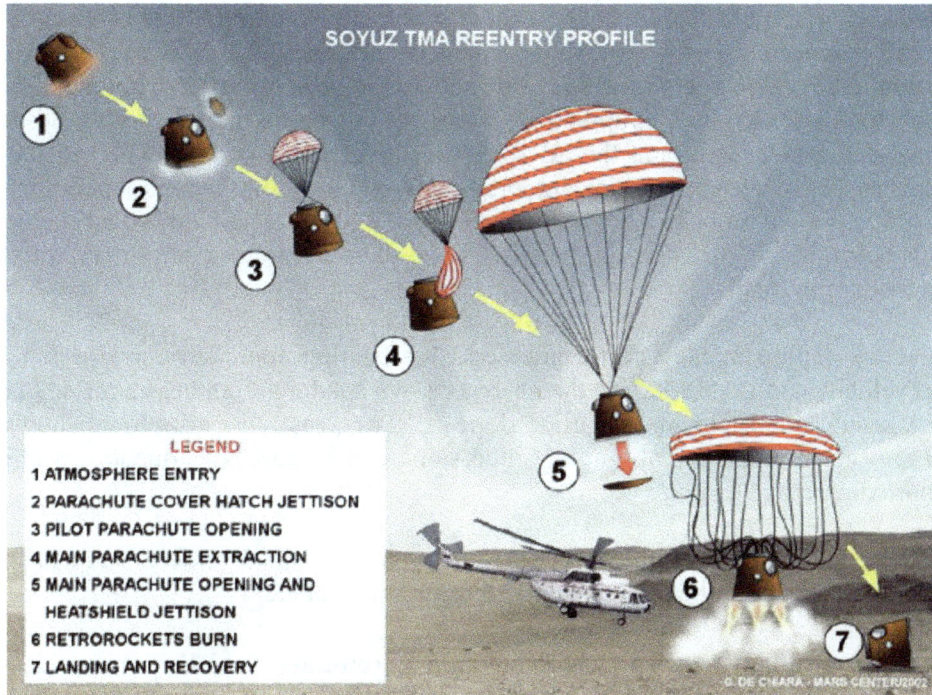

Figure 2.2 Soyuz Re-Entry Profile (credit: G. De Chiara - Mars Center/2002)

Figure 2.3 Soyuz Landing Zone (source: NASA)

Space Shuttle Nominal Landing

The re-entry sequence of the Space Shuttle is largely automated. Before the de-orbit burn is performed, the orbiter is turned to a tail-first attitude so that the Orbital Maneuvering System (OMS) engines can be used to slow the orbiter to permit de-orbit. The Reaction Control System (RCS) thrusters are then used to turn the orbiter back into the nose-first attitude needed to properly dissipate the heat of re-entry.

At an altitude of at about 400,000 ft, the orbiter's velocity is gradually slowed by a series of banks and roll reversals. As the atmospheric density increases, the forward RCS thrusters are turned off, while the aft RCS jets continue to maneuver the orbiter until a dynamic pressure of 10 lb. per square foot is sensed by instruments on board. At this point, the ailerons on orbiter's delta-shaped wings begin to operate and the aft RCS roll thrusters are stopped. When the dynamic pressure reaches 20 lb. per square foot, the orbiter's wing elevators become operational and the RCS pitch thrusters are stopped. A speed brake on the vertical tail opens when the orbiter's velocity falls below Mach 10. Then, at Mach 3.5, the rudder is activated and the final RCS burns -- the yaw jets -- are stopped.

At an altitude of 45,000 ft, the Orbiter intercepts the landing approach corridor at the desired altitude and velocity and is steered into the nearest of two heading alignment circles. At 49,000 ft, the orbiter is in subsonic flight and at a glide slope of 19 degrees. Just prior to touchdown, a flare is required to bring it into its final landing glide slope of 1.5 degrees. Touchdown is made at a speed of approximately 190 knots.

Post-landing operations of the Shuttle Orbiter relies upon specialized vehicles and teams that service the orbiter and assist the crew. Among these teams include:

1. **Self-Contained Atmospheric Protection Ensemble (SCAPE).** This vehicle contains the equipment necessary to support recovery including recovery crew SCAPE suits, liquid air packs, and a crew who assist recovery personnel in suiting-up in protective clothing.

2. **Vapor Dispersal Unit.** This unit produces a directed wind stream of up to 45 mph that can blow away toxic or explosive gases that may occur in or around the orbiter after landing.

3. **Coolant Umbilical Access Vehicle.** This vehicle provides access to the aft port side of the orbiter where ground support crews attach coolant lines from the Orbiter Coolant Transporter.

4. **Orbiter Coolant Transporters.** This unit provides refrigeration through the orbiter's umbilical into its cooling system.

5. **Purge Umbilical Access Vehicle.** This vehicle is similar to the Coolant Umbilical Access Vehicle in that it has an access stairway and platform allowing crews to attach purge air lines to the orbiter on its aft starboard side.

6. **Orbiter Purge Transporter.** This unit carries an air conditioning unit that blows cool or dehumidified air into the payload bay to remove possible residual explosive or toxic gases.

7. **Crew Hatch Access Vehicle.** This vehicle consists of a stairway and white room equipped with special orbiter interface seals with pressurized filtered air to keep toxic or explosive gases, airborne dust or other contaminants from getting into the orbiter during crew egress.

8. **Astronaut Transporter Van.** As its name implies, this van is used to transport the flight crew from the landing area.

9. **Helium Tube Bank.** This unit provides 85,000 cubic feet of helium to purge hydrogen from the orbiter's main engines and lines.

10. **Orbiter Tow Vehicle.** This vehicle is used to move the orbiter from the landing facility to the Orbiter Processing Facility and then to the Vehicle Assembly Building.

11. **Mobile Ground Power Unit.** This unit provides power to the orbiter if the fuel cells have to be shut down.

The main job of the recovery convoy is to service the orbiter, prepare it for towing, assist the crew in leaving the orbiter and finally to tow it to servicing facilities.

After landing, the first staging position of the convoy is 200 ft. up wind from the orbiter. The safety assessment team in the SCAPE van moves to about 100 ft. of the port side of the orbiter. A SCAPE-dressed crew then moves to the rear of the orbiter using a high range flammability vapor detector to obtain vapor level readings and to test for possible explosive hazards and toxic gases. Two readings from three different locations are made to determine concentrations of hydrogen, monomethyl hydrazine (MMH), hydrazine, and ammonia. If they find that high levels of gases are present, and if wind conditions are calm, the Vapor Dispersal Unit -- the mobile wind machine -- moves into place and blows away the potentially dangerous gases.

Meanwhile, the Purge and Coolant Umbilical Access Vehicles are moved behind the orbiter and the safety assessment team continues to determine whether hazardous gases are present in the area. Once the umbilical access vehicles are in position, and as soon as it is possible to connect up to the liquid hydrogen transfer umbilical on the orbiter, the ground half of the on board hydrogen detection sample lines are connected to determine the hydrogen concentration. If the concentration is less than 4 percent, convoy operations continue. However, if it should be greater than 4 percent, an emergency power down of the orbiter is ordered. The flight crew is evacuated from the orbiter immediately and the convoy personnel clear the area and wait for the hydrogen to disperse.

If the hydrogen level is below four percent, the carrier plate for the starboard liquid oxygen transfer umbilical is attached to permit insertion of purge air ducts. After the carrier plates have been installed, the freon line and purge duct connections are completed and the flow of coolant and purge air through the umbilical lines then provides cool and humidified air conditioning to the payload bay and other cavities, removing any residual explosive or toxic fumes.

When it is determined that the area around and in the orbiter is safe, the first priority is to assist the flight crew off the orbiter. The Crew Hatch Access Vehicle moves to the hatch side of the orbiter. Then when the access white room is secured, the orbiter hatch is opened and a physician boards the orbiter to make a brief preliminary medical examination of the crew. The crew then leaves the orbiter and departs in the Astronaut Transporter Van. The ground support team then remains inside the orbiter to make preparations for ground towing operations, installing switch guards and removing data packages from onboard experiments. Finally, after waiting for the orbiter's tires to cool, the Tow Vehicle crew installs the landing gear lock pins, disconnects the nose landing gear drag link. Two hours after landing, the orbiter is towed off the runway.

2.2 Contingency Rescue Operations for Land Landing Spacecraft

All contingency rescue plans are based on the premises established in the Agreement on the Rescue of Astronauts, the Return of Astronauts and the Return of Objects Launched into Outer Space. This agreement was first circulated and signed in 1968, is updated on a regular basis, and remains current to this date. For a spacecraft that has inadvertently landed within the territorial boundaries of a foreign nation, a rescue would need to rely upon the SAR capabilities available within the region that the spacecraft lands in, in accordance with the Space Act of 1967 if the landing site is within the territory of a signatory nation.

2.2.1 Space Act of 1967

An international agreement on the rescue of astronauts was created through consensus of the United Nations General Assembly in December 1967 to elaborate on the rescue provisions detailed in Article V of the 1967 Outer Space Treaty, which state:

> *In carrying on activities in outer space and on celestial bodies, the astronauts of one State Party shall render all possible assistance to the astronauts of other States Parties.*

The Rescue Agreement essentially provides that any state that is a party to the agreement (a 'Contracting Party'), must provide all possible assistance to rescue the personnel of a spacecraft who have landed within that state's territory, whether because of an accident, distress, emergency, or unintended landing. Once within the sovereign territory, rescue operations shall be subject to the direction and control of the Contracting Party, which shall act in close and continuing consultation with the launching authority. Responsibility of the rescue operation is thus transferred to the nation in which the spacecraft lands and the launching authority shall "co-operate with the Contracting Party with a view to the effective conduct of search and rescue operations."

If the distress occurs in international waters or in an area that is beyond the territory of any nation, then any state party that is in a position to do so shall, if necessary, extend assistance in the search and rescue operation. In this case, the agreement simply requires that contracting parties, which are in a position to do so, shall "extend assistance in search and rescue operations for such personnel to assure their speedy rescue" and the authority of the rescue operations remains with the launching authority.

Figure 2.4 shows the nations that have signed and ratified the Space Act of 1967. In the planning of space missions, it would be preferable to reduce the possibility of landing within the territory of a country that has not agreed to the provisions of the Space Act by selecting ascent groundtracks that avoid such areas and landing areas that are sufficiently far from unsigned countries. That said, it is often that physical nature of environments and desirable orbital properties that constrain decisions of launch site location, landing site location, and launch inclination more than political and legal constraints.

Lacking in this provision is a detailed definition of what constitutes an 'astronaut', as the treaty is thus unclear whether this provision applies to space tourists or commercial astronauts in addition to astronauts trained by and flown by government space agencies.

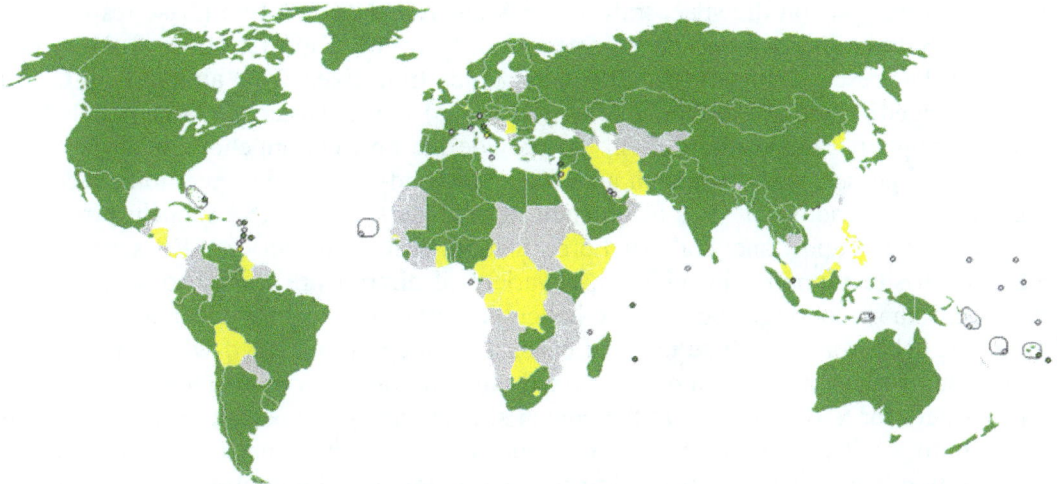

Figure 2.4 Signatory nations of the Space Act of 1967. Green countries have signed the Act and ratified the provisions. Yellow countries have only signed the Act.

2.3 Contingency Rescue Operations for Water Landing Spacecraft

All contingency rescue plans for water landing spacecraft also rely on the Space Act of 1967 and rely upon as many 'layers' of free support as possible, which include globally organized networks to support mariners and aviators in distress. That said, such rescue teams generally rely upon ships of opportunity which rarely have any knowledge as to the operational procedures and hazards involved in spacecraft rescue.

In addition to the resources established in the Space Act provisions, contingency plans for the Space Shuttle and Constellation Programs have also relied on the active support of Canada, the United Kingdom, and Australia. Canada may provide rescue helicopter support for contingency missions extending up the coast from Maine to Newfoundland; the United Kingdom may provide rescue helicopter support for the southern coastal waters of Ireland; and Australia may provide long-range, in-flight refuelable rescue aircraft to support contingencies from their coastal waters, extending into the Indian Ocean. These support agreements have been established by NASA with foreign government through the U.S. State Department.

For human space flight, a contingency rescue operation begins with splashdown at a location other than the nominal landing zone and too far for the nominal recovery forces to accomplish their tasks according to the nominal time line. The primary effort of a contingency rescue operation is for the rescue force is to remove the crew from the spacecraft and place them onboard a helicopter or ship (or render assistance, if on land) within 24 hours of landing. Retrieving the spacecraft itself is secondary and rescue forces will abandon the spacecraft in order to ensure the safety of the astronauts. After splashdown, the astronauts communicate vital information on crew health, vehicle health, and vehicle location for approximately 15 minutes after splashdown via satellite link.

A contingency rescue plan must be responsive enough to deploy to any location in the world where the spacecraft may land and it must be cost effective enough to not be a major cost driver of the total space architecture. A spacecraft that can safely support the crew longer after landing may rely upon a leaner and more cost effective contingency response network, but delays in post-landing drive risk to the crew and post-landing time requirements drive complexity and thus drive the cost and weight margins on the spacecraft. NASA has relied upon a two-phase response to support the Space Shuttle Program and the Constellation Program. The first phase of the rescue operation involves Pararescue Jumpers (PJs) that may deploy from fixed-wing aircraft (e.g. C-130 or C-17s) with Rigged Alternate Method-Zodiac (RAM- Z) boats, large life rafts, and NASA provided Ground Support Equipment (GSE). To provide the best contingency coverage at the most critical mission phases, such as launch and ascent and end-of-mission operations, recovery assets are pre-positioned and/or put on a heightened state of alert. For NASA launches from the Kennedy Space Center, Cape Canaveral Air Force Station, KSC and Patrick AFB serve as the hub of DoD launch site support. In addition, a variety of aircraft are tasked to support pad, launch, and ascent abort contingencies. SAR helicopters are used to help clear the range for launch and are available for immediate medical evacuation support and fixed-wing aircraft help clear the range and, if necessary, augment other fixed-wing aircraft with SAR duties. C-130s are placed on alert solely for SAR missions at the launch site and along the ascent ground track and may even deploy to a loiter location to improve response time in the event of an ascent abort. After orbital insertion, a global plan is needed to respond to any on-orbit contingencies. Two C-17s in the United States and one C-17 in Australia (from the Royal Australian Air Force) are on alert for astronaut rescue in a contingency of this nature.

2.3.1   First Response: Global Response using Parajumpers

The first response of a contingency rescue operation supporting an American space program has involved air-deployed PJ teams. PJ teams have recovered astronauts on Gemini and Apollo missions, and have provided support for Space Shuttle launches and landings. PJ teams also were designated first responders for any contingency landing in the Constellation Program. To assure a response within 24-hours anywhere on the globe, three PJ teams were placed on stand-by. One crewed C-17 would base from the Eastern Seaboard of the United States, a second team would base from Hawaii, and a third C-17, under command of the Royal Australian Air Force, would be on alert in Sydney, Australia.

If a spacecraft were to land in an unplanned location, PJs would fly to the landing site on a fixed-wing aircraft (e.g. C-17) and deploy from the aircraft with Rigging Alternate Method - Zodiac (RAM-Z) inflatable boats and NASA-provided ground support equipment. Once on scene, the PJs inflate the boats and drive to the capsule (Figure 2.5). They attach a sea anchor to slow capsule drift and a stabilization collar to attenuate motion of the capsule and provide a work platform for the PJs. It takes approximately 30 minutes for a PJ team to deploy into the water, inflate the RAM-Z vehicles, and access a hatch of the spacecraft in distress.

When required, the PJs assist astronauts from the capsule and aid their entry into a raft in preparation for subsequent rescue by helicopter or ship. When the astronauts leave the capsule for the life raft, both the astronauts and the PJs must understand the survival advantages and disadvantages of both a fully or partially-suited astronaut. In some cases, the astronauts may be unsuited when PJs arrive on scene; real-time decisions determine the best course of action to ensure astronaut survival. If necessary, the crew can self-egress from the capsule into a life raft carried onboard the capsule in the event that a post-landing contingency does not permit enough time   to   wait   for   PJ assistance.

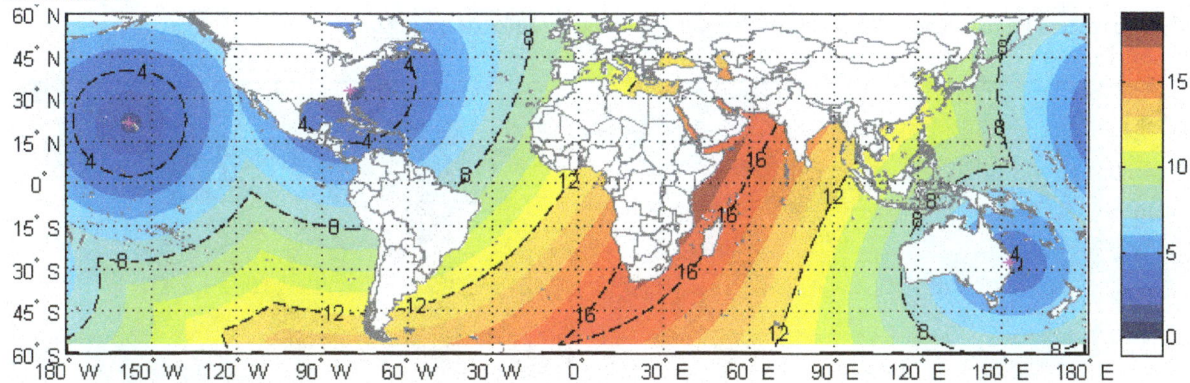

Figure 2.6 shows the anticipated rescue response times supporting a contingency Orion landing at any location on the globe. The response plan assumes three C-17 teams are on alert status: one in Gabreski AFB, NY, one in Hawaii, and one in Australia. The three PJ teams depicted here guarantee a response to anywhere in the world in less than 18 hours, well in advance of the 24-hours the Orion capsule was designed to support its crew post-landing. The Australian plan is a good example of how international agreements can enhance contingency response plans.

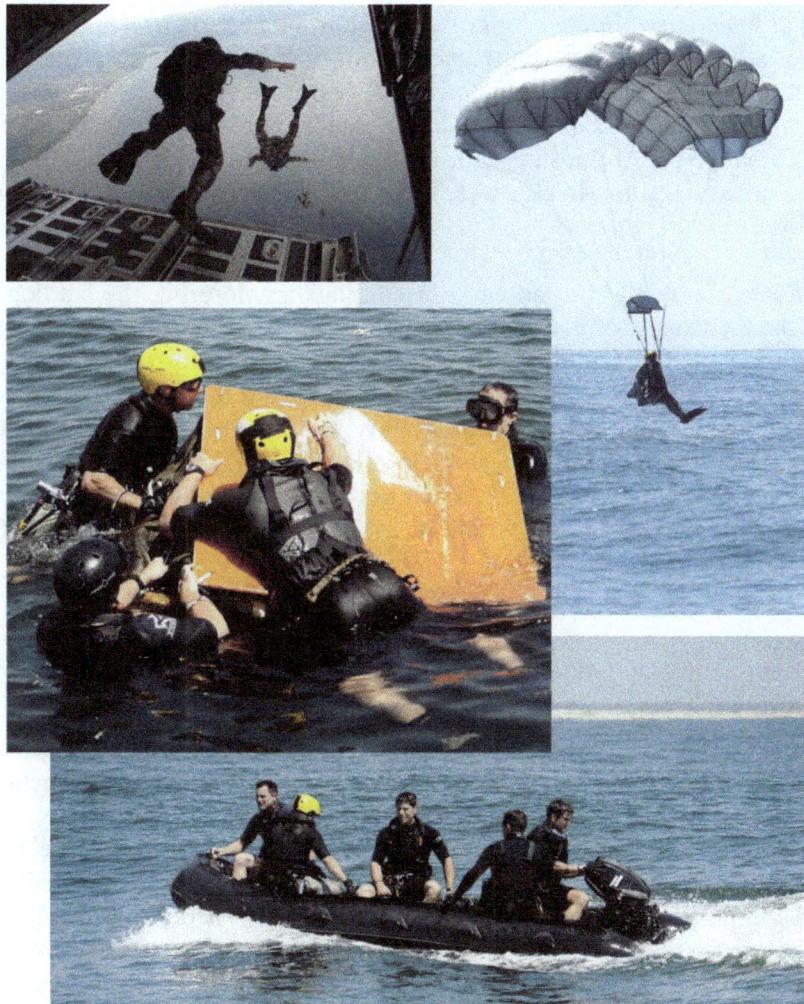

Figure 2.5 Parajumper deployment Sequence (credit: 106rqw.ang.af.mil)

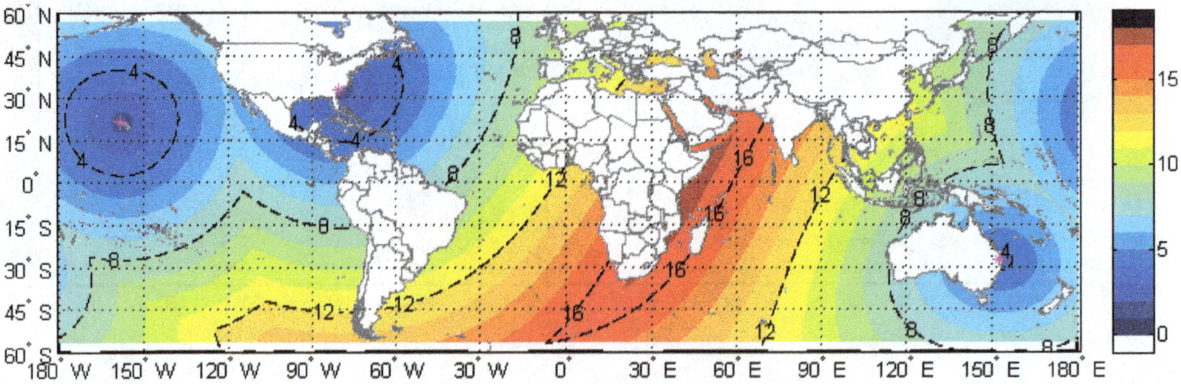

Figure 2.6 First response times for PJ teams (credit: NASA)

2.3.2 Second Phase: Assist using Local Seaborne Assets

The first response will secure the crew on-board RAM-Z inflatable boats, but the rescue operation still needs a second phase that would transfer the astronauts and the PJ rescue team onto a seafaring vessel or helicopter. In the event of a pad abort or an early ascent abort, this vessel may be a helicopter or a boat staged near the launch facility. If helicopters are used to complete the rescue, astronauts are hoisted to the helicopter and, if not previously accomplished, emergency medical care is provided once onboard. If a ship is used to complete the rescue, PJs oversee each astronaut's movement to board the ship and, if not previously accomplished, emergency care is provided once on board the ship. In all cases, PJs oversee astronaut medical care until relieved of that responsibility.

In the event of a late ascent abort or an early or contingency de-orbit, the landing may occur far from any assets the launch site or the primary landing zone may have. For spacecraft designed to land on sea, it is obvious that deploying a network of recovery ships to protect every part of the globe would be cost prohibitive, so the second phase of such a rescue operation would rely upon local SAR capabilities or upon a ship of opportunity that could be diverted to assist in a rescue operation, as it would assist another ship in distress. Fortunately, there already exists internationally-coordinated programs that exist to provide emergency response to mariners and aviators in distress, and these networks would be available to a crew of a spacecraft in distress as well.

2.4 International Program-Specific Agreements

Contingency rescue plans frequently rely upon international agreements to cover areas where there is a significant probability of an emergency landing. An attempted emergency landing by a space shuttle at a site outside of the continental United States, while extremely unlikely, could occur at any time during a flight, and various sites around the world have been designated emergency landing sites through a variety of arrangements between the United States and host governments. An important collaborator to the Space Shuttle Program has been Canada. After it was agreed to construct the International Space Station at an inclination of 51.6 degrees, the ascent groundtrack of the Shuttle would extend from KSC up along the Eastern Seaboard until extending past Newfoundland across the North Atlantic. In the event of multiple engine failures during ascent, the Orbiter may be left too far downrange to execute a Return to Launch Site (RTLS) abort and with insufficient velocity to execute a TAL abort. In this event, the Orbiter would need to land on a suitable runway along the Eastern Seaboard of North America. Prior to the International Space Station program, Space Shuttle missions had generally launched into orbits of lower inclination, so the possibility of aborting to a location on the Eastern Seaboard simply did not exist. The requirements of the Space Station, however, required NASA to seek the support of the Canadians. Fortunately, Canada has always been an enthusiastic supporter of the American Space Program and had agreed to designate certain Canadian airfields as emergency landing sites for the Space Shuttle. These emergency landing sites include:

Gander International Airport, Gander, Newfoundland
Halifax International Airport, Enfield, Nova Scotia
St. John's Airport, St. John's, Newfoundland
Stephenville, Stephenville, Newfoundland
5 Wing Goose Bay, Goose Bay, Newfoundland

Should the NASA Space Shuttle Commander declare an emergency and attempt either an emergency landing at one of these designated Canadian aerodromes, or a 'Mode VIII' crew bailout from the Shuttle while in Canadian airspace or international waters bordering Canadian Territory, one of these Canadian airfields could provide an alternative for a safe landing by the shuttle and its crew. If the Space Shuttle does not land at one of the designated aerodromes, SAR operations will be implemented from the JRCC Halifax in accordance with the *Multilateral Agreement on the Rescue of Astronauts, the Return of Astronauts and the Return of Objects Launched into Outer Space* and Department of National Defense Standard Operating Procedures.

Though the greatest probability of an emergency landing scenario would occur during the ascent phase, there is always a possibility, very remote, that a mechanical malfunction might cause a spacecraft to re-enter prematurely and destined to an unspecified location. In this case, it is critical to avoid the worst of possible scenarios. If the spacecraft was not designed to land on land, then all care must be made to avoid a terrestrial landing. Mountainous terrain, regions of water under excessive sea conditions, extremely remote regions, and land within countries not bound by the International Space Act are all regions that should be avoided. Winged spacecraft have a greater ability to target preferred landing regions, but without parachutes, their landing velocity means that suitable runways are required. Capsule designs have little ability to target regions after being committed to re-entry, but can land in a greater variety of regions.

2.4.1 Global Search and Rescue Plans

The International Maritime Organization (IMO) is the United Nations specialized agency with responsibility for the safety and security of shipping and the prevention of marine pollution by ships. In 1979, the IMO sponsored a SAR Conference to develop provisions of what would become the International Convention on Maritime Search and Rescue. This Conference sought to ensure that SAR services are available everywhere and led to an integrated plan that would urge nations to provide maritime S services for all sea areas and facilitate the exchange of information about the capabilities of each participating nations SAR capabilities. The information collected later became known as the Global SAR Plan or the IMO SAR Plan.

The Global SAR Plan is based upon the international 'Duty to Assist' principle, which states:

"Vessels and aircraft have a duty to provide assistance to other vessels, aircraft or persons in distress, without regard to location, nationality or circumstances'.

This plan assures that SAR services are available wherever mariners might need assistance in a distress situation and that SAR authorities will know where to send or relay distress alerts. To do this, it divides the international waters of the globe into Search and Rescue Regions (SRRs) of which individual countries are responsible. An SRR is an area of defined dimensions associated with a rescue coordination center (RCC), within which SAR services are provided. By definition, there is a one-to-one correlation between SRRs and RCCs.

The IMO is harmonized with aeronautical SAR through the International Civil Aviation Organization (ICAO), which is an agency of the United Nations devoted to aviation safety. Cooperative arrangements between maritime and aeronautical authorities, like cooperation between nations, offer opportunities to provide SAR services more efficiently and effectively. In addition, there are substantial advantages not only to harmonizing aeronautical and maritime SRRs where practicable, but also in establishing joint (maritime and aeronautical) RCCs in some locations.

The IMO maintains a database of the member governments, currently 70, which have provided information on the capabilities of each countries SAR network. This information includes the national responsible authority, the name of each rescue center, contact information, coordination centers for maritime and airborne operations, arrangements for obtaining telemedical advice, SAR facilities available, and the bounds of the region of which each center is responsible.

2.4.2   National Search and Rescue Capabilities

To support the Global SAR plan, individual countries maintain a SAR capability to support mariners and aviators that might be in distress. Presumably, these resources could be relied upon to support spacecraft in distress. In mission planning, the capability of the SAR networks of responsible nations should be considered when identifying potential contingency landing zones. In addition to timely and capable rescue, medical facility proximity and capability should also be considered. Due to their proximity to large areas of ocean, special consideration is often given to the SAR resources in Australia, New Zealand, Chile, and South Africa. Furthermore, since launches to higher inclinations from Kennedy Space Center ascend over the North Atlantic, there is special consideration to SAR resources in Canada, the United Kingdom, and Ireland. Terrestrial landing spacecraft will seek large areas of flatlands to consider for contingency landing sites, and certainly the Kazakh plains and the plains of North America have long been considered throughout the Soyuz Program. A depiction of the global distribution of SRR zones is provided in Figure 2.11.

United States

The US SAR services are provided by a variety of components, the relationships of which are described in the National Search and Rescue Plan that describes the US SAR organization, key authorities and their responsibilities, and its primary principles and policies. The US maintains 11 RCCs: RCC Norfolk, RCC Boston, RCC Miami (which also operates a sub-region from San Juan, Puerto Rico), RCC New Orleans, RCC Cleveland, RCC Alameda, RCC Seattle, RCC Honolulu, RCC Juneau, RCC Langley, and RCC Elmendorf (in Anchorage, AK). There is a one-to-one correlation between each of these RCCs and their associated SRRs, and, to the extent practicable, aeronautical and maritime SRRs are geographically aligned. Maritime regions that fall within the US region of responsibility are coordinated by the Coast Guard, as shown in Figure

2.7 for Pacific Ocean regions and **Error! Reference source not found.** for Atlantic Ocean regions.

The US SAR capability is organized through the US Coast Guard. Using advanced tools such as the Search and Rescue Optimal Planning System (SAROPS) and the manual solution work sheets with manual plotting, the Coast Guard can have a suitable SAR resource ready to proceed within 30 minutes of notification of a distress, leading to a total mission response time to any location within an area of responsibility of no greater than a two hours. SAR coordinators may make SAR agreements with federal, state, local, and private agencies, though international agreements require prior consultation with the Department of State and other interested federal authorities and military commanders may retain control of their own forces conducting SAR for their own units. Federal agencies supporting the US National SAR Plan include the Department of Transportation (DOT), DoD, Department of Commerce (DOC), Federal Communications Commission (FCC), NASA, and the Department of the Interior (DOI).

Federal assets may be provided by the Coast Guard, other Armed Services, the FAA, and the FCC. The Coast Guard maintains a wide variety of SAR resources, primarily dedicated to maritime SAR throughout the U.S. and its territories. Primary resources include long-range HC-130 and medium-range HU-25 fixed-wing aircraft and HH-60, and HH-65 helicopters, High-Endurance Cutters (WHECs), medium-endurance cutters (WMECs), and patrol boats (WPBs), along with buoy tenders, icebreakers, and harbor tugs. US Air Force resources include limited numbers of helicopters and fixed-wing aircraft capable of being used for civil SAR purposes, photographic reconnaissance aircraft and Side-Looking Radar Reconnaissance aircraft, as well as a limited numbers of pararescue personnel that may be available to assist civil resources in specific situations. The Civil Air Patrol (CAP), using corporate and privately owned aircraft, can also be mobilized for search missions over land. US Navy resources include extensive numbers and types of fixed-wing aircraft and helicopters, surface and submarine vessels, sea-air-land (SEAL) teams, diving teams, salvage forces, and radar nets. Naval commanders will normally assist SAR coordinators in handling SAR missions.

In addition, most states have a statewide law enforcement agency that can provide SAR services, though while coordination with state agencies may be important to SAR efforts, operational control of federal resources is not normally assigned to states. County agencies can also provide SAR assistance, as most counties have a sheriff, some with land, water, or air SAR responsibility, and municipal agencies can often provide resources of police and fire departments, as well as medical facilities. Commercial agencies can also be mobilized (e.g. fishing companies, rescue squads, and aviation clubs) as well as private agencies such as the National Association for Search and Rescue (NASAR), Mountain Rescue Association (MRA), and the Diver's Alert Network (DAN).

Canada

In Canada, the Department of National Defense is responsible for providing SAR response in federal or international waters using a team of 750 personnel, which includes ground crew, air crew, and 150 Search and Rescue Technicians, who are advanced trauma life support pararescue specialists. Operations are coordinated through the three Joint Rescue Coordination Centers (JRCCs) that are maintained by the Royal Canadian Air Force (RCAF) and the Canadian Coast Guard (CCG). The three JRCCs are located at Halifax, Trenton, and Victoria. In addition, the CCG staffs a Maritime Rescue Sub-Center (MRSC) in Quebec, though the MRSC in St. John's, Newfoundland was closed in early 2012.

For a spacecraft contingency landing, the RCAF would provide all primary air resources. RCAF squadrons dedicated to SAR operations are stationed at Canadian Forces Base (CFB) Gander, CFB Greenwood, CFB Trenton, CFB Winnipeg, and CFB Comox. Combat support squadrons are also located at CFB Cold Lake, CFB Bagotville, and CFB Goose Bay. These centers, along with their respective regions of responsibility, are shown in Figure 2.8. Secondary air resources could be requested from the Royal Canadian Mounted Police (RCMP), Department of Transport, or the Department of Fisheries and Oceans. The Civil Air Search and Rescue Association (CASARA) supplies civilian aircraft flown by volunteer pilot owners in support of the RCAF's SAR operations upon the request of the RCAF and consists of approximately 375 aircraft and 2596 certified pilots, navigators and spotters.

Seaborne assets are also relied upon to support a spacecraft in distress, and these would be coordinated by the CCG. Similar to the manner in which airborne assets are coordinated by the RCAF, the CCG would provide all primary marine resources in response to a SAR emergency and coordinate additional assets that might be provided by other federal government departments and agencies such as the Royal Canadian Navy (RCN), the RCMP, or the Canadian Coast Guard Auxiliary (CCGA), supplies civilian vessels sailed by volunteer owners.

Figure 2.7 US Regions of Responsibility in the Atlantic (L) and Pacific (R) Oceans.

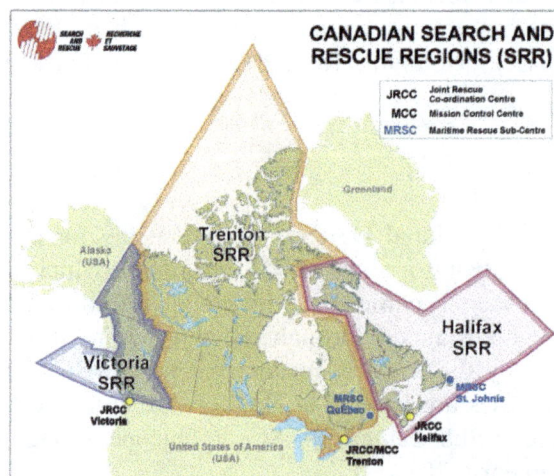

Figure 2.8 Canadian SAR Resource Locations and Regions of Responsibility (credit: Canadian Search and Rescue Secretariat)

Canada has consistently been a key component of spacecraft mission contingency support planning and it has been of critical importance to space missions launching from KSC to the ISS, since the ascent ground track for these missions pass alongside the Canadian Eastern Seaboard before heading east over the North Atlantic. The Halifax JRCC and the resources stationed at CFB Gander play a vital role in the support of these missions. In addition to the support of spacecraft that might land in the coastal waters, the flatlands of Saskatchewan and Manitoba have long been considered contingency landing zones of Soyuz missions, so it is likely that future missions of terrestrial landing spacecraft will continue to consider these areas as viable landing zones.

Australia

Australia has an RCC based in Canberra which is responsible for the national coordination of both maritime and aviation search and rescue within a region of 52.8 million square kilometers of the Indian, Pacific and Southern Oceans. For SAR incidents at great distance from the Australian coast, Australian Defense Force long range assets are normally used, and even with these very capable units, not all of the SAR region can be reached by Australian based aircraft or by ships in a timely manner.

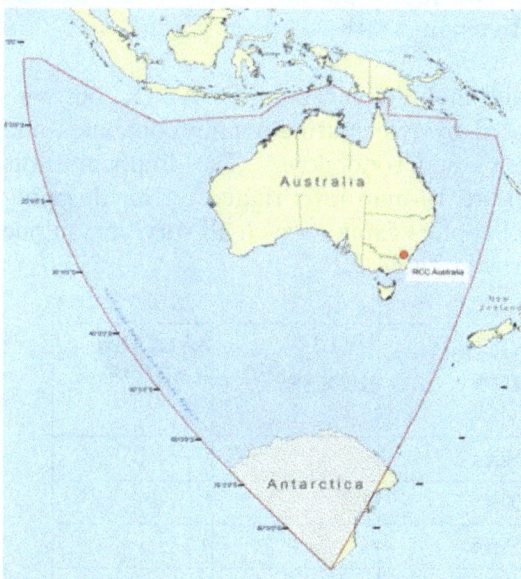

Australia is responsible for a large region of the Southern Hemisphere extending to the South Pole. Because of the capability of the Australian assets and the remoteness of the regions about Australia, it has been a critical component of the contingency response planning of crewed spacecraft missions, and mission contingency response plans often rely upon Australian assets to assist in a potential operation that might occur far beyond the Australian SRR, which was established for the SAR support of aircraft and marine vessels in distress and shown in Figure 2.9.

Figure 2.9 The Australian Search and Rescue Region

RCC Australia, operating from Canberra, coordinates Commonwealth, State and Territory authorities and organizations: The Australian Maritime Safety Authority (AMSA) and the Australian Defense Force (ADF) at the Commonwealth level; and the relevant police service or force at the State and Territory level, are the relevant SAR Authorities within Australia. These organizations can provide airborne and maritime assets organized through a four-tiered system. Tier One has dedicated fixed wing aircraft and crew for supply dropping, homing to beacons and visual search. Tier Two and Three centers have rescue capable helicopters and crew that are already providing emergency medical services and SAR services to the Australian or a

State/Territory Government, that can be engaged by AMSA. Tier Four centers have fixed wing aircraft and crew that can be engaged by AMSA to be used for visual searches. Locations of Tier 1,2, and 3 centers are shown in Figure 2.10.

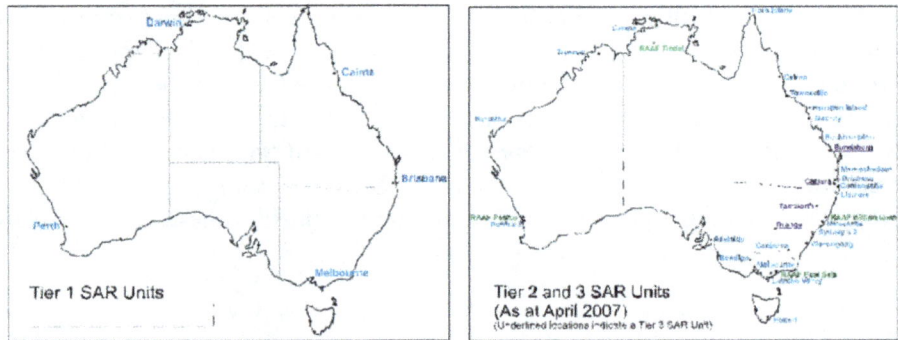

Figure 2.10 Australian SAR Assets by Tier (credit: National Search and Rescue Manual, Australian National Search and Rescue Council)

C-130H and C-130J aircraft are based out of Richmond and deliver a standard five unit Air-Sea Rescue Kit (ASRK), consisting of two x 10 man rafts in separate valises joined by waxed rope to three marine supply containers. Each ASRK contains three marine supply containers, which are packed with stores to suit prevailing requirements, and two radios per raft. DHC4 aircraft based out of Amberly, Townsville, and Perace also deliver a five unit ASRK.

Furthermore, P3C aircraft are stationed at Edinburg and they deliver a similar ASRK, but with four units comprising two 8-man Switlik rafts separated by two marine supply containers and 500m of rope, as well as heliboxes, SAR datum buoys and sonobuoys. The droppable rafts available at East Sale, Pearce, Tindal, and Williamtown are 10-man rafts rigged for air dropping from helicopters, as backup for occasions where the helicopters cannot winch all survivors in one sortie. SAR assets at RAAF bases are summarized below in Table 2.2.

BASE	Air-Sea Rescue Kit (ASRK)	RAM-Z Inflatable Boats	SAR Data Buoys	HELI-BOXES	SALCOM RADIOS
Amberley	4	0	3 Sets	6	0
East Sale	0	2	1 Set	0	0
Edinburgh	8	0	4 Sets	6	16
Pearce	3	2	3 Sets	6	2
Richmond	8	0	6 Sets	6	16
Tindal	0	2	1 Set	0	0
Townsville	2	0	3 Sets	6	2
Williamtown	0	2	1 Set	0	0

Table 2.2 Minimum Requirements for Holdings of Search and Rescue Equipment at RAAF Bases (credit: National Search and Rescue Manual, Australian National Search and Rescue Council)

United Kingdom

The United Kingdom is desirable in space mission contingency planning because of its SAR responsibility to a large region of the North Atlantic. Maritime and Aeronautical Rescue assets are controlled by the modern Aeronautical Rescue Coordination Center at the Royal Air Force (RAF) base at Kinloss, which maintains one Nimrod MPA fixed wing aircraft that can drop liferafts and survival equipment within a range of 800 NM. Other fixed wing aircraft can be requested. Further, the RAF Sea King helicopters possess night vision equipment, thermal imaging capabilities, and medical capability and are on alert at each location with a 15 minute response time are stationed at Boulmer, Chivenor, Leconfield, Lossiemouth, Valley, Wattisham, Culdrose and Prestwick. These helicopters have a maximum endurance of 4.5 to 6 hours and an operational radius of 250 NM and can carry up to 17 passengers.

Kazakhstan

Kazakhstan has had a long history of supporting Soyuz landing operations and will continue to be a viable landing zone for all terrestrial landing spacecraft due to the large areas of high plains than cover its countryside. Kazakhstan's Ministry of Defense currently operates no less than 45 EC145 light-twin helicopters for search and rescue efforts and has recently acquired 20 Eurocopter EC725 helicopters for high altitude or cold weather operations.

Chile

Chile is desirable in space mission contingency planning because it has a long western coast along with adequate SAR and medical facilities that could be targeted by water-landing spacecraft in need of a time-critical de-orbit. Chile is also responsible for a large region of the South Pacific Ocean.

Chile is divided into five SAR districts, each containing a Maritime Rescue Coordination Center (MRCC) that is coordinated by the Air Force of Chile (Fuerza Aerea de Chile), and is responsible for waters off the Chilean cost as well as the Drake Passage extending towards the Antarctic Territory. There is also an Antarctic Naval Patrol placed in this area during summer (from December to March). The MRCCs are located at Iquique, Valparaiso, Talcahuano, Puerto Montt, and Punta Arenas. The Air Force has a pararescue capability that relies upon three C-130 aircraft as well as Bell UH-1 Iroquois, Bell 206B Jet Ranger, Bell 412 and UH-60 Blackhawk helicopters. The Chilean Navy consists of 61 surface vessels based at Valparaiso with maximum speeds ranging from 28 to 30 knots, based at Valparaiso.

South Africa

South Africa is of interest South Africa in space mission contingency planning because of its proximity to large regions of the South Atlantic and South Indian Oceans, but South Africa does not have dedicated SAR units and makes use of assets from government, private and volunteer organizations, so assets are deployed when available. The South African Navy has 12 total ships that have a max cruising speed of 27 knots and normally cruise at 16 knots. They have Super Lynx helicopters onboard with a range of 328 NM. They are stationed at Cape Town and Durban with Reserve Units at East London, Port Elizabeth, Johannesburg, Simon Town and Pretoria. The South African Air Force currently has 175 aircraft, including 9 C-130s stationed at Waterkloof, 10 C-47s, and 77 helicopters. However, these assets are only effective near land; for rescue more than 800 Km from land, the only timely rescue would be a ship in the area.

Ireland

Ireland is desirable in space mission contingency planning because missions originating from KSC bound for the ISS pass over Ireland. A TAL abort of a capsule-based spacecraft for these missions will target waters off the Irish coast. The responsibility of Irish Search and Rescue has been delegated to the Irish Coast Guard, which is responsible for SAR operations within an area that extends 200 NM to the west, 30 NM to the south and to the middle of the Irish Sea. An RCC at Shannon controls three contract helicopters at Dublin, Shannon and Waterford with a forth provided by Aer Corp at Sligo. Other assets can be called into assist and support can be requested from vessels in the vicinity, the Naval Service or from the United Kingdom. The Irish Navy currently has 9 patrol ships stationed at Cork, Dublin, Limerick and Waterford and two Casa CN-235 twin turbo-prop aircraft.

New Zealand

New Zealand is desirable in space mission contingency planning because of its large region of responsibility in the South Pacific Ocean. The Rescue Coordination Center in Wellington responds to 1200 SAR incidents a year and cover 30 million square kilometers from the South Pole to the Equator and from the Tasman Sea to half way to Chile. New Zealand has two frigates in their Navy. Both are stationed on the north shore of Auckland. They can cruise at 27 kts and they have five helicopters that can support them. They also have two fixed wing patrol aircraft.

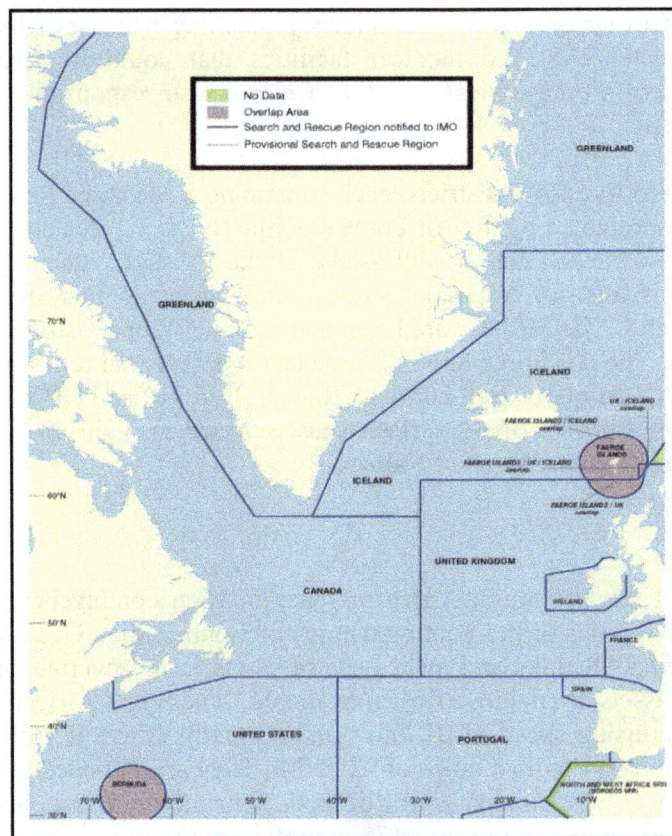

Figure 2.11 Global Search and Rescue (SAR) plan in the North Atlantic

2.4.3   Automated Merchant Vessel Emergency Response

The Automated Mutual Assistance Vessel Rescue (AMVER) System, sponsored by the United States Coast Guard, is a computer-based voluntary global ship reporting system used worldwide by search and rescue authorities to arrange for assistance to persons in distress at sea. AMVER was established in 1958 as an experiment, confined to waters of the North Atlantic Ocean, notorious for icebergs, fog and winter storms, and has expanded to coordinate today over 22,000 ships from hundreds of nations.

The basic premise of AMVER, as a vehicle for mariner to help mariner without regard to nationality, continues to this day. Merchant ship owners or managers enter specific information about their vessels into the AMVER database. Once registered in the system, participating vessels provide a continuous update on their position by sending sailing plans, position, diversion and final (arrival) reports to AMVER. AMVER uses the data to create a central database of vessels and predict their positions at any time. The density plot shown in **Figure** 2.12 shows the way ship traffic is distributed about the globe. In the event of an emergency, an up-to-date register of participating vessels in the area is available for use by RCCs to coordinate a response and divert the best-suited ship or ships as quickly as possible to the distressed vessel or persons at sea.

During the Mercury, Gemini, Apollo and Skylab Programs, AMVER played an important role by providing to NASA a prospective maritime support plan in the event of a space flight emergency. Later, AMVER played an essential part in the contingency rescue plan in support of the Constellation Program and is anticipated to be a key element in protecting the crews of commercial space missions. Figure 2.13 shows the average anticipated response time of the first AMVER ship to a splashdown anywhere within the ISS inclination limits. In approximately 30% of the water surface area, rescue by AMVER ship is likely to take more than 24 hours. These areas are operationally avoided by selection of landing locations on a per-rev basis. Forty minutes of loiter should allow for a water landing site where AMVER support would be expected within 24 hours. Failure scenarios demanding reentry within 40 minutes would be extremely rare.

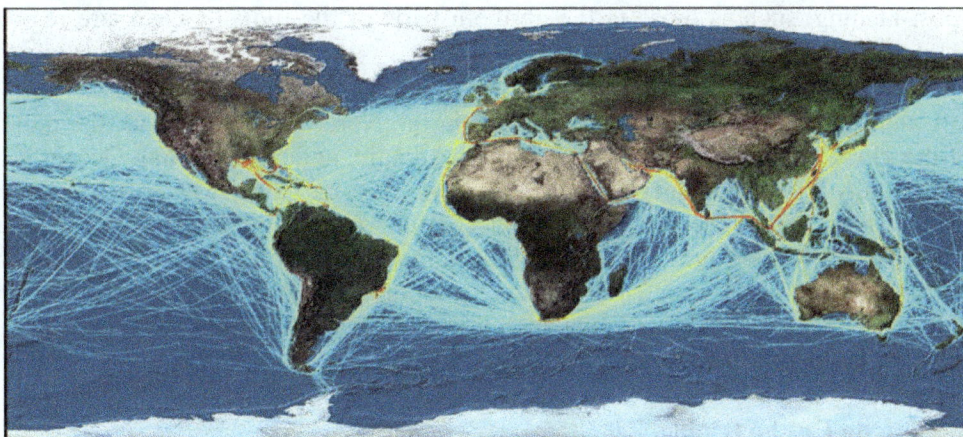

Figure 2.12 AMVER Density Distribution for July 2012 (credit: AMVER)

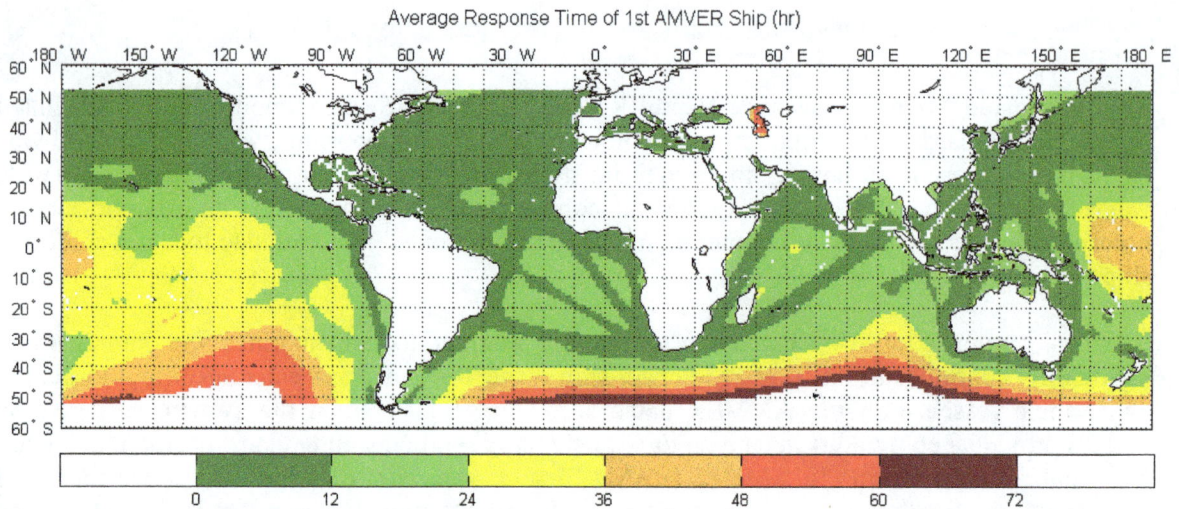

Figure 2.13 Average anticipated response time of the first AMVER ship to a splashdown anywhere within the ISS inclination limits (source: NASA documents)

2.4.4   Landing in Remote Waters

As the primary landing zone is always the most preferable place to land, well-planed missions have contingency landing areas which may be used if weather or other factors make landing at the primary landing zone inadvisable.   Edwards AFB and White Sands Missile Range provided backup landing areas for the Space Shuttle program. There are times, however, when an on-orbit contingency may demand an immediate de-orbit or when a re-entry burn is not executed on time or if the de-orbit burn underperforms grossly.   If neither the primary landing zone nor a contingency landing area can be targeted, a water-landing vehicle would be best landing in an area in proximity to ships or SAR support that could provide a rapid rescue response. Shipping lanes are a preferable region, such as those that extend from North America to Europe or to Hawaii. Conversely, remote areas, regions consistently exposed to excessive sea conditions or bad weather, and regions that have extremely warm water that could drive an emergency egress scenario post-landing, such as those in the northern parts of the Indian Ocean, are regions to be avoided.

As the vast majority of shipping operations occur in the Northern Hemisphere, the most remote regions of the globe are in the South Pacific, South Atlantic, and South Indian Oceans. The first ship of opportunity that might arrive in these waters may be days away, often exceeding the duration of the survival provisions within the spacecraft. An airborne response could be expected well before a ship of opportunity may be available in these regions, as a ship may have to be deployed from a South American, Australian, or South African port.   In addition to these continental ports, there exist few islands in these remote waters that may provide a ship of opportunity to a spacecraft in distress. Of significance are:

1.   **Tristan da Cunha** (United Kingdom)

Tristan da Cunha (37°06′44″S 12°16′56″W) lies in the South Atlantic. There is no airport, though fishing boats from South Africa regularly service the islands. Charter yachts frequent the island in seasons of fairer waters so rescue assets may be possible in the summer months, but less likely in winter months.

2.   South Georgia (United Kingdom)

South Georgia (54°15′S, 36°45′W) also supports a fishing industry, though tourism has become a larger source of income in recent years, with many cruise ships and sailing yachts visiting the islands. There is no airstrip. Charter yacht visits to South Georgia are common in the summer months and last between four and six weeks, though rescue assets are unlikely to be available during winter months.

3.   Chatham Islands (New Zealand)

The Chatham Islands (43°53′S, 176°31′W) reside in the South Pacific and are part of New Zealand. It has a sealed runway at Karewa that supports the Air Chathams airline, which operates twin turboprop Convair 580 aircraft and connects the island to New Zealand daily. Black Robin Freighters also operates shipping services from the Chatham Islands and rescue assets are likely available.

4.   Kerguelan Islands (France)

The Kerguelan Islands of the South Indian Ocean (49°15′S, 69°35′E) support a variety of earth sciences and biology research. There are also some fishing boats and vessels that are owned by fishermen on Réunion Island and some Ukrainian enterprises which may serve as potential rescue assets.

5.   Prince Edward Islands (South Africa)

The only human inhabitants of the Prince Edward Islands (46°46′23″S 37°51′09″E) are the staff of a meteorological and biological research station run by the South African National Antarctic Program on Marion Island. Rescue assets are unlikely.

6.   Bouvet (Norway)

Bouvet Island (54°26′S 03°24′E) lies in the South Atlantic Ocean and is the most remote island in the world. It has no ports or harbors and is difficult to approach. Rescue assets are highly unlikely.

2.5 Contingency Rescue and Recovery Planning

As spaceflight is far from routine, problems can and do occur that would force the crew to abandon the baseline mission, forcing the targeting of an unplanned landing location. Often, such as in the case of an ascent abort or an off-nominal re-entry, rescue forces have insufficient time to adapt to the new landing location. The development of a suitable contingency rescue and recovery architecture is often an iterative process interrelated with vehicle design elements that deal with the post-landing protection of the crew. Usually, post-landing risk mitigation comes at the cost of increased weight penalty in the spacecraft or penalties in the launch availability if a too-conservative threshold of environmental conditions is set.

The post-landing environment can be time critical. For example, the failure of the Command Module Up-righting System (CMUS) system of the Orion spacecraft creates a dire situation in itself if the capsule lands in a Stable Two orientation. The situation quickly becomes life-threatening if the rescue response takes greater than two hours, since the ventilation system will no longer be able to support the crew, and in a Stable Two, the snorkel fan cannot work. Thus, designers must trade the probability of a CMUS failure with the added costs incurred with guaranteeing a two-hour rescue response.

Contingency rescue planning is a much more complicated design problem than nominal rescue planning. Subject to some constraints, contingency landing locations could happen at any location on the globe. Clearly to protect all such locations would be both prohibitive in the cost of the recovery network as well as the in immense safeguards that would have to be integrated into the spacecraft and recovery architecture so that the spacecraft would be able to land and be recovered from virtually any environment. Compromises must be made which consider factors such as overall mission risk, cost, vehicle mass and performance, and operability issues such as the availability of launch and de-orbit windows. To weigh these factors, certain questions must be addressed:

1) What is the probability of initiating a trigger that would result in an off-nominal landing?
2) If a contingency mission profile is entered, where would the spacecraft land?
3) How much accumulative risk would the crew be exposed to in the post-landing phase following a contingency landing?
4) How much vehicle or contingency rescue capability should be provided to protect against such risks and at what cost?

Such questions are rarely addressed within the scope of the providers of rescue and recovery services since they address total mission risk and cost as well as vehicle design, so acceptable risk margins as well as cost of a contingency rescue capability must be put in context with the design trades of the entire space architecture.

2.5.1 Pad and Ascent Aborts

The greatest source of contingency in any space mission happens upon ascent, driven by failure modes primarily associated with the launch vehicle. The Constellation Program estimated the overall probability of an abort trigger on ascent to be 1/451. In comparison, on-orbit failures are roughly two orders of magnitude more rare, and the vast majority of these failures permit sufficient time to target the primary landing zone. The primary objective of the contingency recovery network during the ascent phase is to protect the ascent ground track of the vehicle until orbit is achieved. A global response plan still needs to be maintained for the remainder of the mission, but the probability of occurrence of a contingency response is greatly reduced.

In protecting the ascent phase of a mission, each event in an ascent abort must be weighed – how effective is the system that can detect the problem in sufficient time to trigger the abort, how reliable is the abort, how dangerous is the abort procedure to the crew, how dangerous is the environment the contingency landing will be made, and finally how responsive and effective is the rescue operation. All of these events may be made safer, but only at the expense of vehicle mass or complexity, overall cost, or by imposing stricter operational constraints. Enhancing rescue and recovery architecture can be a large cost driver, so it is important to realize the actual benefit in terms of risk mitigation these enhancements would produce.

As an ascent abort might be triggered anywhere from launch to orbital insertion, inclusive of staging events, main engine cut-off (MECO), and vehicle separation from the launch vehicle. Thus, there is a distribution of probabilistic locations the vehicle might land if an abort is successfully executed. Effective contingency recovery architecture will guarantee a response capable and timely enough to mitigate most post landing risks. Vehicle capability such as post-abort targeting or cross-range capability through the use of wings or other lifting surfaces may be used to consolidate this exposed ground track, thereby simplifying the rescue architecture.

In mission planning, every nominal rescue procedure must be validated through testing in any environment the spacecraft will commit to land in and the spacecraft will not commit to de-orbit if the environments at the intended landing site are in excess of the parameters established through these tests. Contingency rescue operations are a much more complicated. Ascent aborts can result in a wide geographic range of landing sites. Winged spacecraft intrinsically have a cross range capability that permits the consolidation of these ascent abort landing sites to a small number of abort landing sites; the Space Shuttle's abort modes included both an RTLS abort, where the orbiter could separate, turn around, and glide back to a landing at the launch site, and also a TAL abort, where the orbiter could separate and glide to a landing at either Moron, Spain; Banjul, The Gambia; or Ben Guerir, Morocco. If the abort trigger is caused by a failure of more than one of the three main engines, then a contingency abort will result. If the abort is triggered at a time that would not permit a safe glide to an abort landing zone, then the crew would need to egress via a 'Mode VIII' abort and parachute into the ocean. In this case, the orbiter would maintain orbiter integrity for in-flight crew escape and would then be ditched. US Air Force parajumper teams have supported the Space Shuttle program for exactly this possibility – the rescue of the Shuttle crew following a Mode VIII abort.

Contingency rescue planning of winged spacecraft is greatly simplified over those of capsule-based spacecraft. Once abort landing sites have been identified that have sufficient runway length and government approvals, then the weather conditions about the landing zones may be treated as launch commit criteria, where the vehicle would not commit to launch unless the weather at all possible abort landing zones is within the established limits to guarantee a safe landing.

Capsule-based spacecraft have much less capability to target specific zones. Though thrusters on the capsule may provide a downrange or uprange capability to avoid undesirable environments and capsule attitude adjustments may provide a very limited crossrange capability, for the most part the crew is committed to a ballistic trajectory following an abort separation from the launch vehicle. Every ascent abort in a capsule, up to the point where a stable orbit may be achieved, requires a rescue operation. The problem is that one never knows at what time an abort will be triggered, and thus at what location the capsule would splash down. The possible environments into which the capsule could land can be highly varied and spread over a large swath of terrain. To guarantee that every location along the ascent groundtrack would conform to established rescue thresholds may result in a dearth of possible launch opportunities, resulting in a system that is not very operable. For example, if the environmental rescue criteria along all points of the ascent groundtrack were complied with before committing to launch of Constellation's Orion capsule in the winter months, there would only exist about a 20% availability of launch windows from November through April. To improve the availability of launch opportunities, either the vehicle and rescue architecture must be made more robust and responsive or the risk of landing in an excessive environment and driving the overall risk to the crew must be accepted.

So we need to know the failure probabilities that would lead to an abort trigger associated with the launch vehicle and the spacecraft. Further, abort effectiveness is a key data source so that only survivable aborts are considered. Abort effectiveness is a function of the mode of abort, which would consider whether a launch escape tower or a SM engine is used, as well as its environment, which would consider the speed and attitude of the vehicle, the atmospheric conditions and thermal environment, etc. Thus, abort effectiveness varies with mission elapsed time.

Launch trajectories also need to be considered. Capsule-based spacecraft are generally ballistic following an ascent abort, so trajectory extrapolations along each point of the ascent trajectory need be performed. Since launch windows have an opening and closing time, there will be a range of inclinations from the launch site that would need to be modeled, which means that the groundtrack will vary. If the launch could occur anytime during the launch window, then the entire area bounded by the window-open and the window-close trajectories must be considered for capsule-based spacecraft. For winged spacecraft, the situation is simplified. Here, the launch windows are often confined by the accessibility of ascent abort landing facilities along the ascent trajectory; one would not commit a winged spacecraft to launch into an orbital inclination that would extend its flight path beyond gliding distance of suitable landing facilities in the event of an ascent abort.

The trajectory extrapolations following the separation of the spacecraft from the launch vehicle upon ascent for any type of spacecraft may be influenced by a wide variety of dispersions, including variation in the abort motor performance, navigation and re-entry attitude error, and environmental variations. The higher the altitude of the vehicle at time of separation, the greater the exposure will be to dispersions and the greater effect they will have upon the spacecraft's eventual landing location. As dispersions are random in nature, they are generally modeled through Monte-Carlo techniques. As capsule-based spacecraft have historically had a main engine integral to its SM, prograde or retrograde thrusting operations may be used to avoid undesirable landing locations, though the planning of the windows where a prograde or retrograde thruster firing must be defined precisely. The times chosen will influence both the geographic regions where a capsule may land as well as the probabilities that the capsule will land in each region.

Figure 2.14 shows the probability distribution of where the Orion spacecraft may land following an ascent abort. The range of areas where an abort splashdown is possible is bounded by the trajectories that would follow a launch at the launch window open time and the launch window close time. Once the vehicle has obtained sufficient velocity as it approaches the Newfoundland region of the Eastern Seaboard, it can abort to the Shannon, Ireland landing zone or, with a few more seconds of thrust, it can abort to orbit. This figure shows that most of the risk is retired at or near the launch site, but there are still significant risks of an abort trigger that would place the spacecraft anywhere along the Eastern Seaboard region.

Figure 2.15 considers the post-landing response times that could be achieved by a candidate contingency rescue plan. In this figure, color gradients show probability distribution of splashdown locations based off of Ares failure modes (0 to 95.2%) where probability is the percentage, per abort and per launch, of the first response within the exposure time bin. We see here that, with an emergency airborne response originating from Kennedy Space Center and/or Gabreski Air Force Base, a response of less than four hours can be guaranteed for over 90% of launch aborts. If aircraft and aircrews were additionally placed in Shannon, Ireland and Newfoundland, Canada, the guaranteed response time could be 100%, but like everything in engineering, this benefit comes with a cost.

Aborts using the launch escape system were dramatically demonstrated with the spectacular abort of Soyuz 10-1, where a fire that broke out on the launch vehicle led to a successful pad abort seconds before the launch vehicle exploded violently upon the pad. The Space Shuttle program logged four RSLS aborts. Pad and early ascent contingencies in the Constellation Program, as with Apollo, were protected by a launch escape system. By far, the most dominant failure modes of the Constellation Project were integral to the Ares launch vehicle, though the Orion spacecraft itself did have some failure modes that could trigger an abort. The leading failure modes that would trigger a launch abort system (or Mode I) abort were associated with the first stage ignition or the staging event, and most significantly through a failure of the SM fairing to jettison or a failure of the main engines to prime or pressurize.

After a successful staging is confirmed, the launch escape system is jettisoned as the rapidly increasing vehicle velocity reduces its effectiveness. A SM abort (Mode II) would result if an abort is triggered after the launch escape system has been jettisoned. Here, the SM would be used to push the vehicle away from the launch vehicle. A SM abort was successfully executed during the flight of Soyuz 18-1, where the sole cosmonaut was ejected with his Soyuz spacecraft late in the ascent trajectory when the SM failed to separate cleanly. He landed in mountainous terrain in what is now Eastern Kazakhstan. (see Section **Error! Reference source not found.**).

Following a Mode II abort, the SM motor could then be used to push the capsule farther downrange, or turn around and drive the vehicle back uprange, to target a preferred landing location, or at least to avoid areas such as the North Atlantic that could pose hazards to the crew upon and after landing. Such targeted aborts were termed 'Mode III' aborts. A failure in the systems that would manage a Mode III abort would cause the vehicle to downgrade to a Mode II abort and follow a ballistic trajectory to landing. The dominant causes of a failure of the SM to effectively target a preferred landing location were deemed to be a failure of the SM main motor to ignite or a failure of the helium pressurant system or a failure of either of the isolation valves protecting the reserves of the hypergolic propellants in the SM. The Orion SM was designed to use not only the main engine but all eight of its auxiliary thrusters in optimizing its downrange or uprange capability in a Mode III abort. Failures of any of these motors would degrade the overall capability to avoid regions of land or rough water.

To reduce the risk associated with a Mode III abort, the vehicle capability may be enhanced to better avoid undesirable regions or better endure off-nominal landings, or the rescue architecture could be enhanced. Vehicle capability often comes with a penalty in vehicle complexity, mass, and cost. A rescue architecture enhancement often comes with large penalties in cost, and there may be limits on what types of operations may be safely performed. But the capability of a Mode III abort is often confined by the ability of the crew to endure the extended G-forces of an abort maneuver, and 6G Gx limits are generally established.

Following orbital insertion, the Orion spacecraft separates from the upper stage, though its ability to maintain itself in orbit is immediately dependent upon its ability to deploy its solar arrays. A failure to perform this would force the crew to commit to an Abort Once Around (AOA), or Mode IV abort, where the spacecraft would de-orbit using battery power reserves roughly twenty minutes after insertion and land at a location off the western coast of Baja California. The principle failure modes leading to a failure of the solar arrays to deploy were integral to the deployment mechanisms and included a failure of the unfurling motor or the canted hinge.

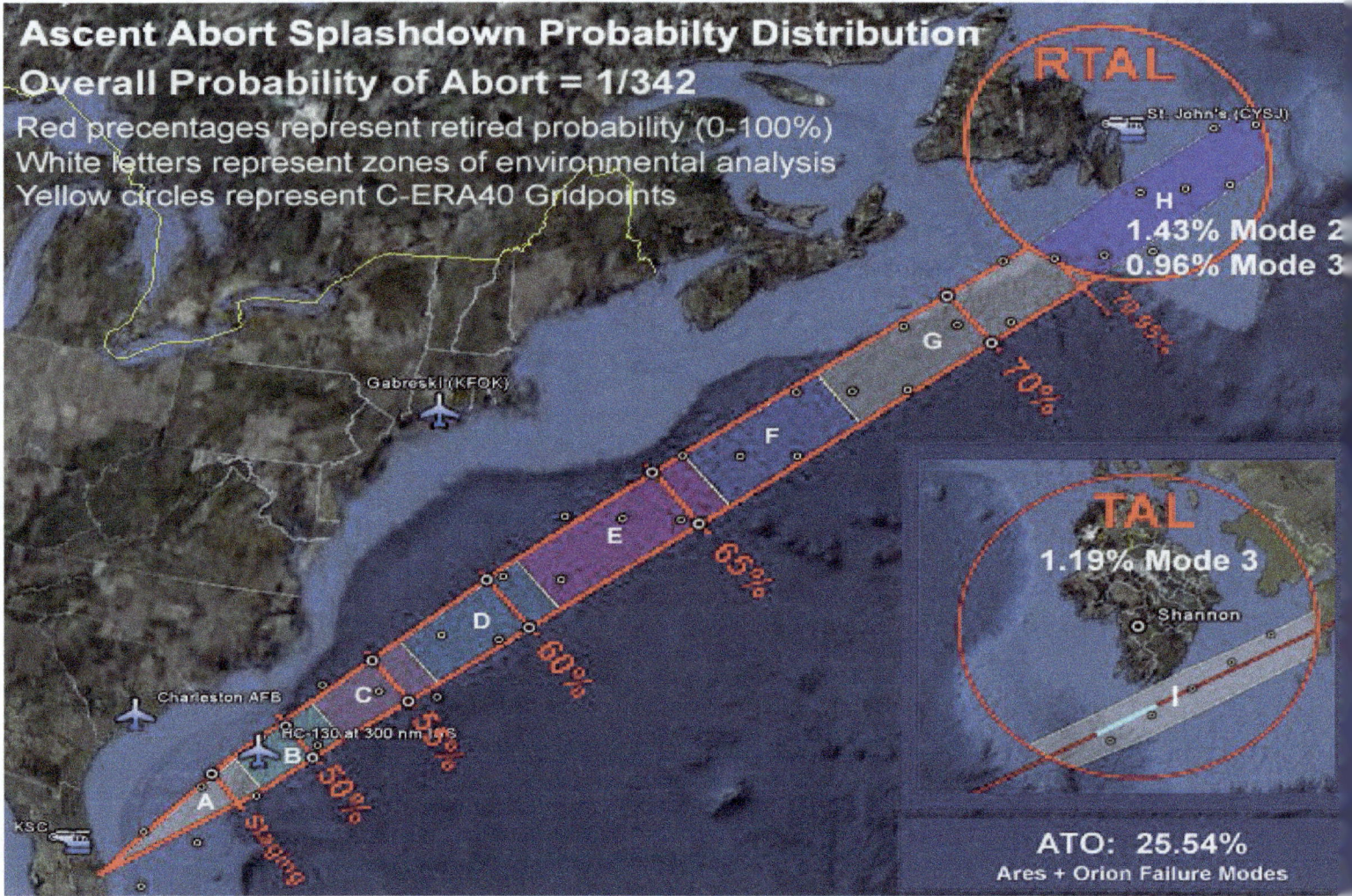

Figure 2.14 Constellation Project Ascent Abort Splashdown Probability Distribution (author's models)

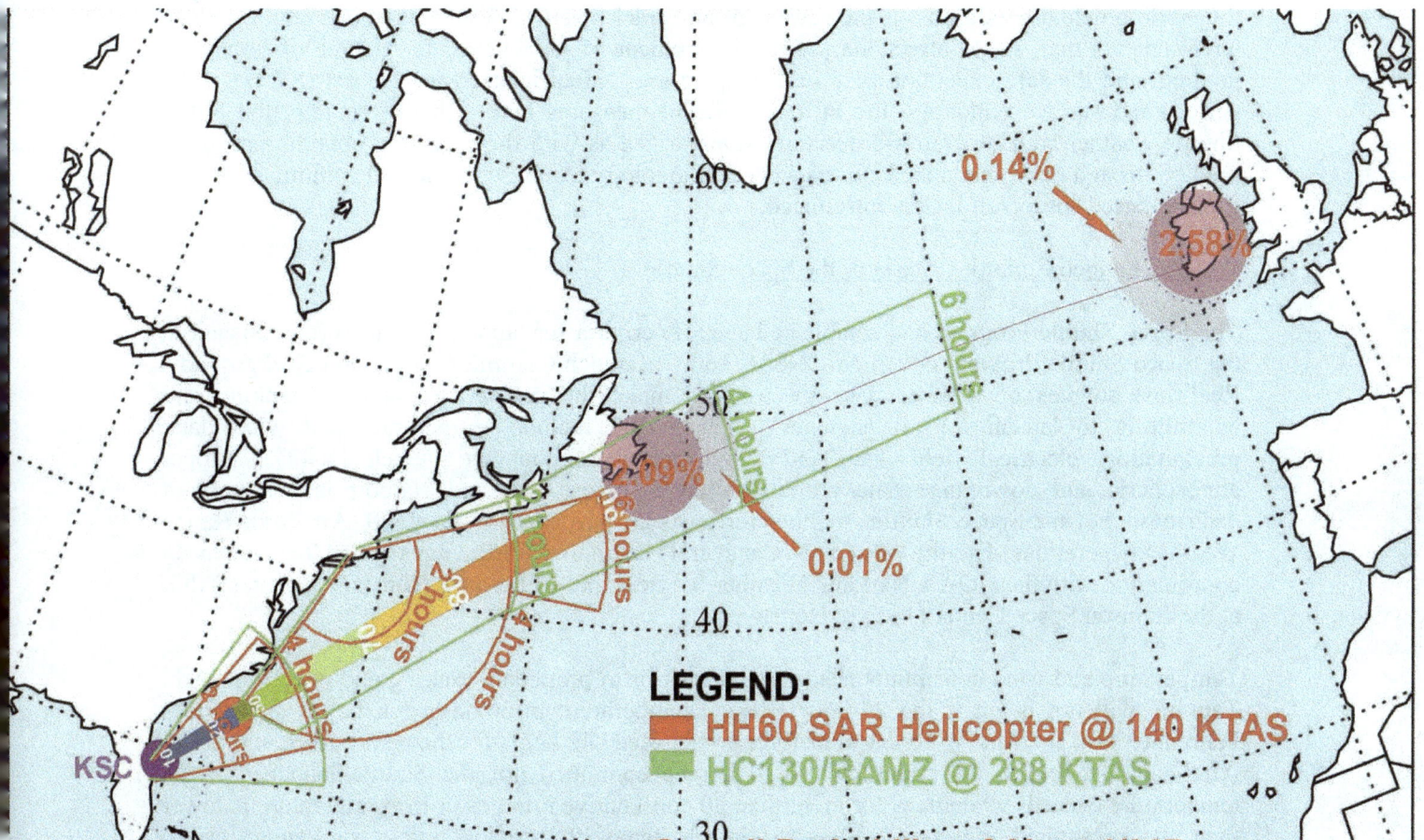

Figure 2.15: Constellation Splashdown Probability Distribution with Recovery Times (author's models)

Operational Constraints

Operational constraints are the limits of environmental parameters into which the vehicle might be flown, below which the vehicle has demonstrated the capability to safely operate and land in, and rescue teams have demonstrated a reliable capability to conduct rescue operations in. Such constraints typically include weather conditions about the launch pad, such as lightning potential, cloud cover, surface winds, and winds aloft, as well as weather conditions about ascent abort landing zones. Capsule-based spacecraft, in addition, may have more stringent constraints over the environmental conditions of the ascent ground track itself. These constraints could include environments that would affect the proper deployment of parachutes, the degree of impact of landing, and the safe execution of a rescue operation. Significant Wave Height (SWH), wave period, and surface winds are the most significant measures here. In this section, the Space Shuttle weather launch commit criteria are summarized as with the means of obtaining accurate predictions and real-time launch day data. A methodology to assess the launch commit criteria of capsule-based spacecraft is then introduced.

Weather Launch Commit Criteria of the Space Shuttle

The Space Shuttle Program had established specific criteria for launch, which were published in the Space Shuttle Program document "NSTS 16007: Launch Commit Criteria and Background". Real-time sources of weather data are used to make the critical decisions of tanking and committing to launch for all launches. These data include temperature and wind data, precipitation, electrical field data, and cloud presence about the launch facility. Upper atmospheric and downrange data are also used to assess risk and support launch commit decisions. For the Space Shuttle, weather forecasts are provided by the U. S. Air Force Range Weather Operations Facility at Cape Canaveral beginning three days prior to launch in coordination with the NOAA National Weather Service Space Flight Meteorology Group (SMG) at the Johnson Space Center (JSC) in Houston.

Temperature and wind constraints need to be met prior to propellant tanking and prior to launch. Tanking will not begin if the 24-hour average temperature prior has been below 41 degrees Fahrenheit or if the wind is observed or forecast to exceed 42 knots for the next three hour period. After tanking begins, the countdown shall not be continued nor the Shuttle launched if the temperature exceeds 99 degrees for more than 30 consecutive minutes or the temperature is lower than the prescribed minimum value for longer than 30 minutes unless sun angle, wind, temperature and relative humidity conditions permit recovery. For launch, the peak wind speed allowable is 30 knots, though the threshold may be lower for southerly winds. Upper atmosphere winds are determined as well through wind balloon releases from Cape Canaveral Air Station.

Precipitation and the potential for lightning must also be considered. Low latitude spaceports, such as Kennedy Space Center, are desirable from an orbital dynamics perspective, but offer unique challenges associated with cumulous activity, especially in the summer months. A launch must be free of any precipitation at the launch pad or within the flight path. Tanking will not begin if there is forecast to be greater than a 20% chance of lightning within five nautical miles of the launch pad during the first hour of tanking. Shuttle launches are constrained if lightning has been detected within 10nm of the pad or the planned flight path within 30 minutes prior to launch. Electrical field measurements are also consistently made to reduce the risk of a lightning strike of the vehicle upon ascent, such as had happened to the Apollo 12 spacecraft.

The presence of clouds constrain launch opportunities as well. Convective clouds, especially those that can build electrical fields, pose the greatest danger to spacecraft and launch is prohibited through cumulus type clouds with tops higher than the 41°F isotherm level, within five nautical miles of the nearest edge of cumulus type clouds with tops higher than the 14°F level, within ten nautical miles of the nearest edge of cumulus clouds with tops higher than the -4°F. level, or through an observed thunderstorm, cumulonimbus debris cloud, or anvil cloud. Launch may occur, however, if the cloud layer is a cirrus-like cloud at temperatures below -5°F that shows no evidence of containing water droplets. The space shuttle will also not launch if any part of the planned flight path is through a layer of clouds any part of which is within 5nm is 4,500 feet thick or greater and the temperature of any part of the layer is between 32° and -4°F.

The immediate downrange environments must be protected as well. Cloud ceiling and visibility thresholds must be met. For the Space Shuttle, direct visual observation using optical tracking or a forward observer on ascent is required through 8,000 feet, though certain exceptions may be made for the presence of clouds at lower altitudes provided that the vehicle integrity may be observed through 8,000 feet.

Finally, landing zones that might be used in the event of an ascent abort must have favorable weather conditions. For the RTLS and TAL sites, cloud coverage 4/8 or less below 5,000 feet and a visibility of 5 statute miles or greater, with no detached opaque thunderstorm anvils less than three hours old within 15nm of the runway, or within 5nm of the final approach path extending outward to 30 nautical miles from the end of the runway, is required, though an exception may be made for rain showers at TAL sites if specific conditions are met. For AOA and PLS sites, cloud coverage must be 4/8 or less below 8,000 feet and cloud coverage criteria are slightly less stringent. For all sites, the crosswind component may not exceed 15 knots with headwinds not to exceed 25 knots and tailwinds not to exceed 10 knots average with no more than a moderate level of turbulence.

Weather Launch Commit Criteria of Capsule-Based Spacecraft

Winged spacecraft, due to their cross-range capability, have the ability to target specific landing sites following an ascent abort. The Space Shuttle, for example, can return to a landing strip by the launch site through an RTLS abort or press ahead to a landing at a TAL site. As long as the weather is within established thresholds at these sites, a launch will be go. Capsule-based spacecraft, however, do not have such capability. Though adjustments in the attitude of the capsule may provide a small amount of downrange or cross-range capability, an ascent abort of a capsule-based spacecraft will result in a ballistic trajectory to landing. Thus, while a winged spacecraft may consolidate the areas where the vehicle may land following an ascent abort into several contingency landing zones, a capsule-based spacecraft could land anywhere along the ascent groundtrack up to the point where orbital insertion could be obtained, and thus much larger geographical regions must be protected.

An ascent abort of a capsule-based spacecraft will result in a rescue operation, but since one never knows prior to launch where and when the abort will be triggered, one never knows into what environments the capsule might land. Operational thresholds are established as criteria for deployment of the rescue personnel since each procedure of a rescue operation involves rescue personnel that volunteer to perform a potentially hazardous mission and the lives and the safety of these personnel must be considered along with the safety of the spacecraft crew in need of rescue.

For a capsule in water, one needs to know 1) the SWH, 2) the Sea Surface Mean Wind Speed, and 3) the average wave period. The SWH and the average wave period contribute to the waters dynamic motion and thus the oscillatory behavior of the capsule in the water. Excessive sea conditions may pose safety hazards for the rescue personnel and will create a hazardous situation for the crew in case an unassisted egress need be performed. Furthermore, as rescue personnel have historically been deployed from aircraft via parachute, a surface wind threshold must be established for safe deployment of the PJs. Preliminary resting of the Orion capsule via the Post-Landing Orion Recovery Test (PORT) test established these thresholds at a 5m SWH for rescue operations and a 13.9 m/s surface wind velocity for PJ deployment.

In addition to the environments immediately about the capsule, the aircraft to be deployed must also not be confined by operational constraints, which would include being free from environments that could be prejudicial to the safe operation of the aircraft, such as takeoff and landing weather minimums, excessive icing conditions, or excessive convective activity. PJs must also deploy with adequate visibility and into water temperatures compatible with the temperature ranges of their immersion suits. Every operational procedure must be validated through testing in an analog environment and the skills of the rescue personnel must be kept current through recurrency training.

As it is unlikely that there will be many opportunities where sea conditions and winds are within operational thresholds along the entire ascent groundtrack, a compromise must be made. It may be acceptable, for example, to commit a vehicle to launch knowing that there is a 5% chance that the vehicle might land in excessive sea or atmospheric conditions. This is a trade that balances the operability of the spacecraft architecture with the overall mission risk. Clever use of 'Mode III' targeting can help minimize exposure to such conditions. If the vehicle were to land in excessive conditions, decisions must be made whether to wait out the environments or commit the crew or rescue teams to a hazardous rescue operation.

For a capsule with a terrestrial landing system, rescue operations are greatly facilitated since the crew would probably be in a better situation to conduct a safe self-egress. But land terrain is far less predictable than ocean terrain; Soviet cosmonauts Vasili Lazarev and Oleg Makarov learned this abruptly after the ascent abort of their Soyuz 18-1 spacecraft resulted in the craft tumbling down a mountainous snow-covered slope and almost over a 152m precipice, were it not for the parachute lines entangling themselves upon vegetation.

Obtaining real-time environmental data

Though weather predictions are useful to support decisions of tanking and launch preparation, real-time environments data is essential to support a decision to commit the vehicle to launch. Satellite data can be problematic as launch windows would need to be coordinated with satellite overpasses, but ground based meteorological stations, remote-sensed data, and airborne observations provide essential information of the weather environments about the launch site, the ascent groundtrack, and any contingency landing zones.

The weather equipment currently used by the forecasters to develop the launch and landing forecasts in support of the Space Shuttle Program is:

1. Radar (Patrick AFB) - Provides rain intensity and cloud top information out to a distance as far as 200 nautical miles.

2. NOAA National Weather Service NEXRAD Doppler Radar (Melbourne, FL) - Provides rain intensity and cloud top information out to a distance as far as 200 nautical miles. The NEXRAD radar can also provide estimates of total rainfall and radial wind velocities.

3. Launch Pad Lightning Warning System at Cape Canaveral Air Station (CCAS) - Provides data on lightning activity and surface electric fields induced by charge aloft.

4. Cloud to Ground Lightning Surveillance System (CGLSS) (KSC) - Detects and plots cloud to ground lightning strikes within 125 nautical miles of the Kennedy Space Center.

5. Lightning Detection And Ranging (LDAR) (KSC) - Helps determine the beginning and end of lightning conditions.

6. National Lightning Detection Network (US National Network) – Provides a broader assessment of lightning conditions from the launch site.

7. Rawinsondes (KSC) – Provides temperature, dewpoint and humidity, wind speed and direction, and pressure data up to altitudes of 100,000 feet.

8. Jimsphere Baloons (KSC) - Provides highly accurate information on wind speed and wind direction up to 60,000 feet.

9. Doppler Radar Wind Profiler (KSC) - Measures upper level wind speed and direction over Kennedy Space Center from approximately 10,000 feet to 60,000 feet to ensure conditions have not changed significantly since the last data collection from a Jimsphere balloon.

10. Rocketsonde (KSC) - Senses and transmits data on temperature, wind speed and direction, wind shear, pressure, and air density at altitudes between 65,000 feet and 370,000 feet, if needed.

11. Satellite Imagery (Global Data Set) – Provide cloud cover data obtained directly by the geostationary GOES weather satellites.

12. 33 Meterological Towers (KSC, CCAS, and the Shuttle Landing Facility) – Provide real-time wind, temperature, and atmospheric moisture data.

13. Buoys (Ascent Groundtrack) – Provide hourly measurements via satellite of temperature, wind speed and direction, barometric pressure, precipitation, sea water temperature, and wave height and period along key points of the ascent groundtrack.

In addition to data received from the sources listed above, real-time data is also related from the solid rocket booster retrieval ships, positioned 150 nautical miles downrange, and often a T-38 aircraft being flown by a weather support astronaut. At Mission Control, all of these weather data are integrated onto a single display, called the Meteorological Interactive Data Display System (MIDDS). Data received from landing sites that might be used in the event of an ascent abort are also displayed. If all weather conditions are within limits at the launch site and all ascent abort landing zones, the Space Shuttle is committed to launch.

For capsule-based spacecraft, the environmental conditions along all points of the ascent groundtrack become much more important. Downrange radar systems and/or real-time data obtained from airborne platforms can be used to provide a distribution of environments along large swaths of sea, such as SWH and wave period, and also characterize critical atmospheric data that could affect a landing, such as wind shear, or a rescue operation, such as surface winds and cloud layer minimums. Such data can be integrated and averaged to provide support to launch commit decisions.

Environmental Modeling Techniques

The environment about the spacecraft must always be considered for both the nominal mission as well as any credible contingency scenarios. While considering launch availability in the design of spaceflight architecture, historical data is needed to best predict the environments into which a spacecraft may abort into. These data are essential to best select vehicle performance requirements, spaceport characteristics, and baseline reference missions. While making launch day decisions to commit to launch or a de-orbit burn, real-time data is essential to accurately quantify the risk that the crew would face during launch and ascent. Thus, environmental models help best design efficient space architecture while sources of real-time environmental data best support operational decisions of launch or de-orbit.

Winged spacecraft can target landing strips where environments may be determined through a history of Meteorological Aerodrome Reports, or METAR. These reports provide real-time information on winds, temperatures, dew point, pressure, and cloud cover – data that would be essential before committing to land. In modeling the typical availability of any landing zone protecting a spacecraft ascent, historical data may be averaged at all contingency landing zones and weighed with the probability that the spacecraft might be forced to land at each zone. On launch day, real-time data is provided to the Flight Director from each potential landing zone before the vehicle is committed to launch. Similarly, landing zones may be such characterized before committing to a de-orbit burn.

Capsule spacecraft are more complicated since the dispersion of potential landing zones may encompass the majority of the ascent groundtrack. Models of atmospheric conditions as well as terrain models, if the spacecraft ascends over land, and models of sea conditions, if the spacecraft ascends over sea, are essential. If a capsule spacecraft ascends over land, the areas of the ascent groundtrack that would be either hazardous to the crew to land on or hazardous to people or sensitive ecological areas must be identified and appropriate means to avoid such areas must be integrated into the spacecraft abort system. Steep terrain, such as mountain ridges, is generally hazardous. Bodies of water, if the spacecraft is not designed to land in water, may also be considered hazardous. Populated areas need to be avoided as well. All of these areas may be avoided by implementing enhanced downrange, uprange, or cross-range capability to the spacecraft or by placing constraints on the inclinations to which the spacecraft launches.

Capsule spacecraft avoid such constraints by launching over water. This explains why spaceports at equatorial latitudes and on the Eastern shores of land masses are most desirable; such locations optimize the use of the Earth's angular momentum as low latitudes allow for direct insertion into any inclination and prograde (Eastern) launches launch with the Earth's rotation. Further, waters about NASA's Kennedy Space Center and the European Space Center's spaceport in French Guiana are relatively benign and there are few islands that may pose constraints. Thus, once ship and air traffic are cleared around the launch area, only the natural environments of the atmosphere and the sea need be considered.

In the design of a capsule-based space system that intends to launch over water, both atmospheric and sea condition models must be used. Atmospheric conditions may prevent the parachute systems from functioning correctly or prevent the ability of parajumper rescue teams to deploy. Excessive sea conditions may prevent a safe rescue of the crew, prevent the crew from conducting a safe emergency egress, or prevent the deployment of rescue teams.

One of the most established global atmosphere and surface condition model is the ERA-40 model. The ERA-40 model has characterized environmental conditions over the period from September 1957 through August 2002 through the use of radiosondes, balloons, aircraft, buoys, satellites, and scatterometers and interpolated to a 40km resolution. Data from the ERA-40 model is averaged monthly, enabling forecasts for the time range spanning 10 to 30 days, a time frame that is short enough that the atmosphere retains some memory of its initial state and long enough that the ocean variability has an impact on the atmospheric circulation, and provides seasonal trends and patterns of variability of significant wave height in the Northern Hemisphere during the last 40 years in terms of means, 90th and 99th percentiles, and return value estimates. [Wang and Swail, 2004]

The ERA-40 set evolved from a series of global environmental analyses. The first data set, produced at the American National Center for Environmental Prediction and the National Centers for Atmospheric Research (NCEP/NCAR) in 2001 used global wind data to force the generation ODGP2 spectral ocean wave model (Cox and Swail, 2001) onto a global $1.25° \times 2.5°$ latitude/longitude grid. This model was enhanced with improved North Atlantic wind data in 2002 to produce wave fields on a $0.625° \times 0.833°$ latitude/longitude grid in the North Atlantic and later in 2004 by the European Centre for Medium-Range Weather Forecasts (ECMWF) to produce a global $1.5° \times 1.5°$ latitude/longitude grid, which became the ERA-40 dataset.

A distinguishing feature of ECMWF's model is its coupling to a third generation wave model, the well-known WAM (Komen et al., 1994), which makes wave data a natural output of ERA-40. The WAM model is a global wave model that is based on first principles that was developed prior to launch of a series of next generation ocean remote-sensing satellites in the 1990s. WAM evolved from wave models developed in the 1960s and early 1970s and specifically addressed the problem of modeling the complete energy balance of the ocean, a problem that was circumvented through assumptions in earlier models. The vast majority of the dataset is freely available through the website: http://data.ecmwf.int/data/.

In the years following its development, the ERA-40 dataset has been extensively validated against observations and it was found that the ERA-40 dataset modeled the averaged dynamics of the ocean more accurately than prior models but severely underestimated high sea states and displayed inhomogeneities in time due to the assimilation of different altimeter sea state data. To address these issues, the ERA-40 model was corrected through use of a nonparametric regression method where data from locations common to ERA-40 measurements and measurements from the Topex satellite were available. The result was a new 45-year global 6-hourly dataset - the Corrected ERA-40, or C-ERA40, dataset. The C-ERA40 is the model currently used by NASA to model global environments and help establish constraints.

2.5.2 Early De-orbit Scenarios

In order to determine the credibility of a landing in a location outside of the PLZ once a stable orbit has been achieved, one needs to know what events might occur that would cause the crew to intentionally de-orbit the spacecraft at a time where it would land in a contingency landing zone or elsewhere. Unlike unintentional failures associated with the actual entry and descent procedures which may cause the spacecraft to land in a location outside of the PLZ, early de-orbit maneuvers are an intentional operation designed to bring the crew into safety from an event that would exaggerated by remaining in the orbital environment. There are two types of scenarios that could require a de-orbit at a time that might not allow for a landing at the PLZ: mechanical scenarios and medical scenarios.

Mechanical Scenarios Requiring De-orbit

Due to the redundancy intrinsic to spacecraft, many mechanical failures may be successfully managed without the abandonment of any mission objectives. Some failures, however, may lead to a re-prioritization of mission objectives. The gravest failures, such as the oxygen tank explosion on the Apollo 13 spacecraft, force the mission to abandon its objectives and devote all of its resources to the safe return of the crew.

For spacecraft in a stable Earth orbit, certain mechanical failures may demand a timely de-orbit, though it may still be prudent to de-orbit at a time when favorable landing conditions and a timely rescue response would be available. The magnitude of the failure must be assessed and the time the crew would have before the situation grew critical must be determined. A critical situation could be caused by the depletion of life sustaining consumables on the spacecraft, a depletion of electrical power, or the depletion of propellants required to de-orbit the spacecraft.

Table 2.3 shows an assessment of conditions considered by the Constellation Program. Propulsion leaks, such as a leaking propulsion tank seal or solenoid, are the most difficult to quantify but place the crew in a 'use or lose' situation where the crew must de-orbit if the leak could not be isolated. Astronauts Neil Armstrong and David Scott experienced a stuck thruster on the Gemini 8 mission, which created a pitching and roll rate that threatened to drive them to lose consciousness. Certainly a de-orbit burn can not be performed if the propellant leak causes the spacecraft to enter uncontrolled attitude fluctuations, so a controlled de-orbit resulting from a leaking propulsion system assumes a stabilized spacecraft.

The ability of the crew to access space suits in orbit greatly reduces the risk that a failure that a vital Environmental Control and Life Support System (ECLSS) failure would drive an immediate de-orbit, but if the crew is committed to wearing space suits as life support, it is likely that a de-orbit would be performed at the next good opportunity. A failure in the coolant loop, however, could drive an immediate (less than one orbit) de-orbit scenario.

Finally, a loss of the spacecraft to receive or store electrical power could be a cause for imminent de-orbit. The Orion spacecraft is dependent on a timely deployment of the solar array within its first orbit. A failure to initially deploy the solar arrays would force the crew to de-orbit within the timeframe of one orbit, landing off the Western Coast of Baja California in an AOA abort. The failure of any system essential to providing and maintaining charge to the vehicle's batteries, and failure of the batteries themselves, is cause for a de-orbit within one orbit, since electrical power is essential to performing the de-orbit operation, maintaining attitude, deploying the drogue and main parachutes, establishing communications and beacons, and driving the up-righting bag pumps. The failure of any of these systems would jeopardize the lives of the crew.

Medical Scenarios Requiring De-orbit

Like mechanical failures, a medical emergency can drive an urgent need to terminate a mission. NASA defines a loss-of-crew (LOC) as the serious injury or fatality of at least one crew member. A serious threat to the health of any crew member will force the mission to abandon any objectives aside from the quick care of the injured crew member.

Medical events that could happen on the ISS are grouped into three categories, each requiring a different response plan. Class I medical events are minor and have no associated mission impacts. A Class II medical event is a significant medical which may or may not impact the mission. Class II medical events may be manageable so that the mission may proceed as planned (a Class IIa event), may require the affected crew member to return at the next available opportunity for further evaluation and treatment (a Class IIb event), or may be manageable but may necessitate an emergency evacuation if the condition worsens (a Class IIc event). A Class III event requires an immediate emergency evacuation. The time-critical nature of Class III medical events may lead to the consideration of secondary or contingency landing zones, if response and medical capability are adequate to handle the emergency.

Spaceflight Morbidity affecting Mission Performance and Safety

Mercury 9 (May 16, 1963)	Elevated CO_2 levels and loss of power to control system, requiring manual reentry.
Soyuz 10-1 (April 23 1971)	Air supply to DM became contaminated post-landing, causing the cosmonaut to lose consciousness.
Apollo 17 (December 1972)	Back strain
Apollo-Soyuz Test Project (1975)	Apollo crewmembers exposed to nitrogen tetroxide contamination upon landing and subsequently developed chemical pneumonia.
Soyuz 21 (August 24, 1976)	Crewmember illness linked to Environmental Control System problem.
Salyut 7 (November 11, 1982)	Kidney Stone resolved on-orbit
Soyuz T-13 (September 1985)	Hypothermia and CO_2 toxicity
Salyut 7 (November 21, 1985)	Crewmember fell ill with prostatitis and urosepsis requiring early termination 56 days into a 216-day mission.
Mir 2 (1987)	Crew member developed tachy-dysrhythmia during an EVA and returned early on next mission of opportunity.
STS-40 (June 1991)	Mechanical malfunction caused formaldehyde toxicity and headaches.
Mir 18 (1995)	Crewmember experienced asymptomatic, sustained ventricular tachycardia. No mission impact.
Mir 18 (1995)	Crew member suffered traumatic eye injury which was resolved on-orbit.

| Mir 23 (February 23, 1997) | Fire introduced smoke and toxic fumes in the station, causing mild second-degree burns. Treatment given on-orbit |
| Mir 24 (February 1998) | Three crewmembers exposed to elevated amounts of CO_2, producing headaches. |

Summarizing these incidents, there have been four traumatic events which were all treated on-orbit. A total of nine cardiopulmonary events included three dysrhythmias of which one event required an early crew member return. Three events involved internal medicine, including one crew return due to a chronic headache. Finally, four events are attributable to gastrointestinal maladies, including one case of prostatitis that resulted in an early return. In all three cases where a medical event necessitated an early return, none were time-critical enough to consider the use of a secondary or contingency landing zone.

Medical Emergencies in Antarctica

Due to their remoteness and relative inaccessibility, Antarctic research stations are often used as good analogs to space habitats. Evacuation capability is often very limited and can be non-existent for periods of over eight months due to the extreme weather of an Antarctic winter. Aircraft may be unable to land and ships may not be able to access coastal bases due to the extent of the winter sea ice. Crews destined to winter-over at Antarctic bases are screened medically to a similar extent as those destined for spaceflight.

The US operates three permanent Antarctic bases: McMurdo and Palmer near the Antarctic coastline and Amundsen-Scott at the geographic South Pole. Since 1954, these stations have averaged a total of 1200 occupants during the summer months and 125 throughout the winter, with one fatality occurring each year, on average. The incidence of medical evacuations from McMurdo has been calculated to be 3.55 evacuations per operational month. Since the medical capabilities of McMurdo Station are comparable to those at the ISS, this statistic can serve as a good proxy. When considering weather and operational constraints, an evacuation can require up to 48 hours transporting an ill or injured person from McMurdo to Christchurch, New Zealand. Transport from Amundsen-Scott can require significantly more time. It should be noted, however, that medical facilities would exist on an Antarctic evacuation aircraft where none would exist on a Soyuz spacecraft.

From 1992 through 1996, there have been a total of 71 medical evacuations from McMurdo. 34 of these have been attributable to trauma, eight from cardiopulmonary problems, seven from dental emergencies, six from internal medicine problems, five requiring breast disorders or gynecological problems, four involving genito-urological issues, three for psychiatric issues, two for surgical issues, and two for neurological issues.

Soyuz 21

On August 24, 1976, the Soyuz 21 mission was truncated for reason of medical evacuation. Cosmonaut Vitali Zholobov was reported to have suffered an illness apparently caused by nitric acid fumes which had leaked from the Space Station Salyut's propellant tanks. Other reports, however, indicate that the crew failed to follow their physical exercise program and suffered from lack of sleep. But sources at NASA have reported that psychologists with the Russian Aviation and Space Agency cited Soyuz 21 as ending prematurely due to unspecified "interpersonal issues" with the crew.

Because Soyuz 21 was returning early, it was outside the normal recovery window. It then encountered strong crosswinds as it descended, causing uneven firing of the retrorockets and leading it to make a hard landing 200 km (120 mi) from the PLZ. Soyuz 21 showed that, even if the PLZ may be targeted in an off-nominal mission plan, the emergency may create urgency where the vehicle could be committed into landing environments less than ideal.

Modeling Early Returns due to Medical Emergencies

Table 2.3 identifies the most relevant medical issues that would drive a need for early termination of a mission, based upon a recent consensus of aerospace medical experts at JSC. The most severe and likely conditions are highlighted in red. These issues are imminently life threatening and create an urgent need to land, with the most probable condition being a gastrointestinal (GI) bleed. A GI bleed can be mitigated partly by intravenous fluid (IVF) injection, if available on the spacecraft, though de-orbit at the first Primary Landing Zone (PLZ) opportunity would still be essential. Similarly, an acute pulmonary embolism or shortness of breath can be stabilized if resources of oxygen are available to the affected crewmember and decompression sickness (DCS) can be potentially mitigated on-orbit. Regardless, all of these scenarios require the crew to de-orbit at the first safe opportunity. It is worth noting, however, that none of the medical conditions studied were urgent enough to consider targeting a location other than the PLZ. In all cases, it is far better to arrive at a location with optimal recovery and medical capability, even if more time in orbit is necessary.

Condition	Likelihood	Severity	Time to De-orbit
Propulsion: Leaking SM tank	Extremely low	Potentially severe	Depends on leak rate
Propulsion: Leaking He tank, Solenoid, or Latch Valve	Extremely low	Potentially severe	Depends on leak rate
ECLSS: ATCS Freon Coolant Loop failure	Extremely low due to redundancy	Potentially severe	<1 rev
ECLSS Cabin Fire	Very low. Sustained fire negligible	Can be mitigated by having crew in suits	>1 rev. Can stay in suits for 5 days
ECLSS: Cabin Pressure Integrity	Very low. Sustained fire negligible	Can be mitigated by having crew in suits	>1 rev. Can stay in suits for 5 days
EPS: Solar Array failure	Significant	Moderate to Severe	< 1 rev
EPS: SM or CM MBSU failure	Extremely low due to redundancy	Moderate to Severe	< 1 rev
EPS: CM battery failure	Very Low	Can be mitigated	> 1 rev
Mechanisms: Solar Array Deploy Failure	Significant	Moderate to Severe	< 1 rev
LRS: Uprighting system bottle pressure	Extremely Low	Moderate	TBD

Table 2.3 Orion Mechanical Failures driving a forced De-Orbit (author's model with help from T. Walker)

Condition	Likelihood	Severity	Time to De-orbit
Gastrointestinal Bleeding	Possible	Severe	Based on IVF available (about 7 liters). ASAP (Next PLZ)
Acute Myocardial Infarction (Heart Attack)	Unlikely due to screening	Moderate to severe	Next PLZ
Acute Pulmonary Embolism / Shortness of Breath	Unlikely	Severe	ASAP (Next PLZ) 24+ hours if O2 supply adequate
Decompression Sickness (DCS)	Unlikely	Moderate to severe	Next PLZ
Neuro symptoms (stroke, DCS)	Unlikely	Moderate to severe	Next PLZ
Renal stones	Possible	Moderate (severe if obstructed)	24 + hours
Appendicitis	Possible	Moderate	24 to 48 hours
Cholelithiasis/ cholecystitis	Unlikely	Moderate	24 to 48 hours
Burns	Possible	Moderate to severe	24 hours. If severe – next PLZ
Smoke inhalation	Possible	Moderate to severe	Next PLZ
Eye injury / globe puncture	Possible	Moderate	24 to 48 hours
Abdominal pain NOS (diverticulitis, obstruction)	Possible	Moderate	24 to 48 hours

Table 2.4 Orion Medical Emergencies driving a forced De-Orbit (credit: M. Chandler)

Soyuz 18-1 was intended to send a crew of two consisting of Commander Vasili Lazarev and flight engineer Oleg Makarov to the Salyut 4 space station for a 60-day mission. The launch and ascent proceeded nominally until late in the ascent profile, at T+288.6s and an altitude of 145km, when the second stage was to separate. Only three of the six pyrotechnic bolts that retained the second stage activated, leaving the massive second stage still attached when the third stage's engine ignited. Fortunately, the thrust of the third stage was enough to eventually sever the remaining bolts and free itself from the deadweight of the second stage, but not until the trajectory had been altered enough to initiate the automated abort sequence. The Soyuz spacecraft separated, the orbital and service modules were discarded, and the descent module committed itself to a reentry.

The Soyuz abort profile was designed to produce peak accelerations in excess of 15g's, but the initial attitude and altitude of Soyuz 18-1 exposed the crew to peak accelerative forces of 21.3 g's! The Soyuz abort system was designed for survivability, but placed the crew at serious risk of injury. It was later determined that that Lazarev had sustained injuries resulting from these excessive forces. But despite the extreme re-entry, the landing system functioned as it was designed, the parachutes opened properly, and a nominal landing was made near the town of Gorno-Altaisk, 829 km north of the Chinese border, and along a steep snow-covered slope. Immediately, the DM began to roll towards a 152m precipice. Were it not for the parachutes becoming snagged on vegetation, the crew would have fallen under their collapsed parachutes to their deaths, but as fate would have had it, the crew was able to egress the DM safely. Prepared for the local temperature of −7 °C, the crew donned their cold-weather survival clothing and promptly destroyed any sensitive materials, fearing that they had landed in China, which was in poor relations with the Soviet Union at the time. The deep snow, the high altitude, and the terrain complicated the rescue effort and the crew was not air lifted out until the following day.

Lazarev and Makarov were in a perilous location on a mountain slope and facing dropping temperatures, but the situation was deemed stable enough to wait for morning when the weather would improve. The crew was adequately equipped safety of the rescue team had to be weighed, so the decision was made to delay the rescue until the next day. Later, the crew of Soyuz 23 was almost lost as rescue efforts prolonged and hypothermia and CO_2 toxicity became extreme. Careful timing and consideration of the environments can be the difference between life and death in such rescue operations.

Figure 2.16 Oleg Makarov (left) and Vasili Lazarev (right) of Soyuz 18-1 (credit: Roskosmos)

2.5.3   Contingency Landing Zones

The Primary Landing Zone (PLZ) is always the most preferable location in which to land; trained rescue teams, ideal recovery assets, well characterized environments, and proximity to adequate and well-staffed medical facilities are available there. For winged spacecraft, the PLZ can be reached with more frequency than with a capsule-based spacecraft of less cross-range capability. During its short free flight period during re-entry, Orion had a cross-range capability of only 44 km (25 nmi) from the nominal ground track. A capsule-based spacecraft needs to wait before the precession of the groundtrack allows for a de-orbit to the PLZ, and these opportunities may take many revolutions and be subject to the environmental conditions at the landing site. For Orion, landing opportunities at the PLZ occur on average once every two days, though this time may be shortened through the use of phasing maneuvers which can be used to adjust the projected groundtrack through fine adjustments in the period of the orbit. If weather conditions preclude a landing attempt, the landing must be rescheduled, which might take several days if no phasing maneuvers are employed. If, however, a de-orbit must be performed to mitigate a serious mechanical or medical problem, the primary landing zone may not be an option. In this case, a contingency landing zone must be targeted. Contingency landing zones are selected in advance for each orbit so that, in the event an emergency would require a rapid de-orbit, a contingency plan would already be in effect.

If a spacecraft has docking capability with the ISS, the crew would have several available options. The first option would be to fly to, or back to, the ISS and dock. This would grant the resources available on the ISS until the next landing opportunity presents itself. Another option would be to remain in free flight until the next landing opportunity at the PLZ. Yet a third option would be to re-enter prior to a landing opportunity at the PLZ, targeting instead a contingency landing zone.

The Soyuz Program relies upon large areas of flat land to assure a safe landing. Such terrain is certainly less common than bodies of water, so landing opportunities are much rarer. Further, as Soyuz has no real cross range capability, it is much more vulnerable to on-orbit contingencies that permit little time to de-orbit. Since these are rare, it is likely that the Soyuz program accepted these risks in light of the benefits of landing on terrestrial plains. The Soyuz capsule had available three emergency landing zones: 1) the Sea of Okhotsk, 2) the steppes of Kazakhstan, and 3) the plains of the United States. Figure 2.17 shows the landing zones that could have been used for both the Soyuz 19 and Soyuz-33 Missions. The landing zones in North America include areas within Manitoba, Saskatchewan, North Dakota, Texas and, Oklahoma. The Texas landing point for Soyuz-33 was quite close to Dallas!

The Constellation Program, like Apollo, sought the benefit of a high degree of flexibility of landing areas with the simplicity and weight savings of a water-nominal spacecraft, but at the expense of more complex and costly rescue operations. For a sea-nominal capsule dependent upon a recovery ship, the decision to land at an opportunity earlier than the next nominal opportunity is limited by the recovery ship speed, which is approximately 15-20 knots. To minimize the time required before a landing opportunity would be available at the PLZ, both the orbit of the spacecraft could be altered through phasing maneuvers and the recovery ship could relocate to the new targeted site.

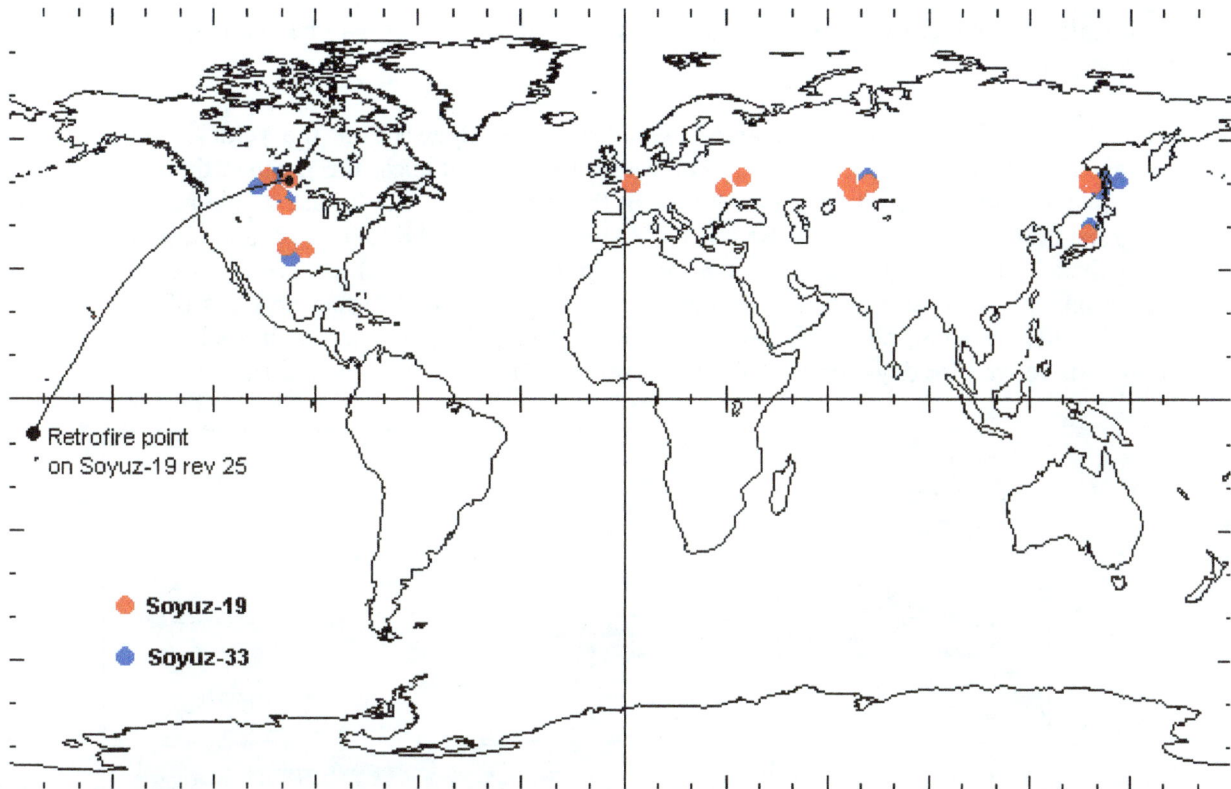

Figure 2.17 Contingency Landing Zones of the Soyuz 19 and Soyuz 33 Missions (NASA documents)

Figure 2.18 shows the preferred landing areas considered by the Constellation Program in the event the PLZ could not be targeted. The blue lines and areas enclosed by blue circles are considered desirable. As the groundtrack progresses from the west to east in a prograde orbit, areas of water near a western coast of a landmass that have a history of benign sea conditions were identified first. Regions near countries with strong SAR capabilities and medical facilities were especially desirable. Regions with a history of rough seas, such as the southern Chilean coast and the South African coast, were excluded. Further, since the Design Reference Missions of the Constellation Program never included orbital inclinations greater than that of the ISS, no areas higher than a latitude of 51.6 degrees were considered, since the groundtrack would never extend to higher latitudes.

Also considered as viable landing zones were areas that have historically demonstrated a high volume of sea traffic, such as the shipping lanes connecting the Mediterranean Sea and major North American ports, the Western United States and Hawaii, Hawaii and major East Asian seaports, and sea lanes headed from the Cape of Good Hope to Indonesian and Southeast Asian ports. Such shipping lanes show a high probability that an AMVER-registered ship of opportunity could arrive on scene within approximately 12 hours of splashdown.

In the extremely rare case that a de-orbit must be performed in less time than would be needed to arrive at a contingency landing zone, then it just becomes important to make sure the spacecraft does not land in particularly hazardous areas. Unlike American capsule-based spacecraft, all regions of water are considered hazardous to the crews of Soyuz capsules. As with Apollo, all land surfaces were considered by Constellation to be hazardous, if not fatal, to the crew.

Politically unstable areas are undesirable and mountainous terrain is to be avoided at all costs for any program.

An orbital vehicle should be designed considering the longest time the vehicle could overfly terrain prejudicial to a safe landing. For water-nominal vehicles, the greatest time exposure to land that a spacecraft orbiting at the altitude of the ISS capsule would face would be 38 minutes – a groundtrack spanning from South Africa through the Middle East and emerging from Kamchatka. Once excluding land, the next regions to be avoided are areas of sea with rough environments, such as the North Atlantic or the Cape of Good Hope, or areas of very remote oceans, such as the South Pacific or South Indian Oceans. A crew landing in such remote regions may need to wait for days for a sea faring vessel to complete the rescue effort.

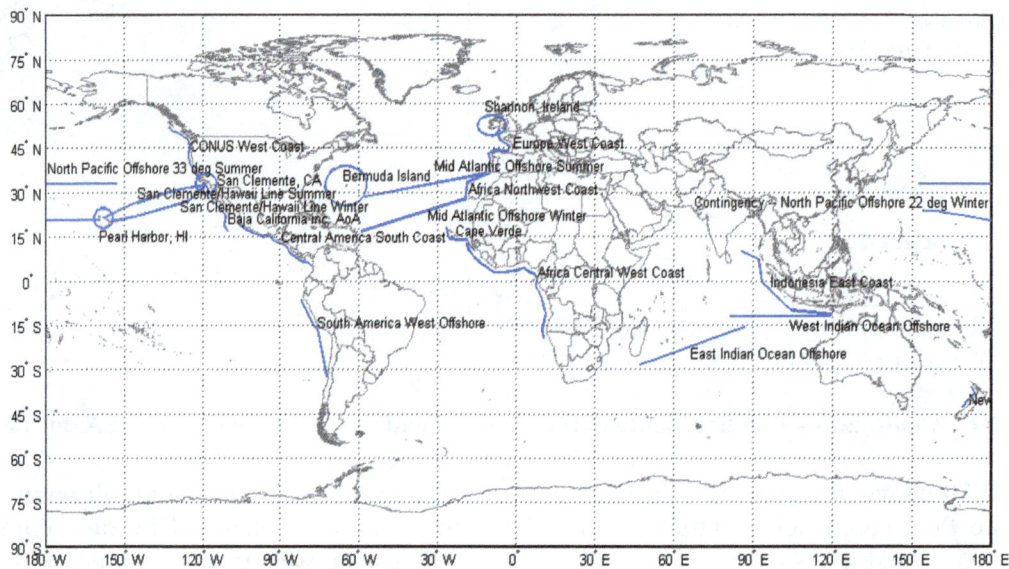

Figure 2.18 Constellation Contingency Landing Zones (credit: NASA)

2.5.4   Erroneous Re-entry

Re-entry failures can alter the trajectory significantly, but generally not in way that cannot be sufficiently covered by the deployed recovery team at the primary landing zone. Ballistic re-entry profiles can cause the spacecraft to land several hundred kilometers short of the intended landing zone. Re-entry error, such as that on Scott Carpenters MA-7 flight can cause similar error as well. To date, no contingency landing areas have been targeted, though alternate landing zones are frequently targeted due to weather conditions. Alternate landing zones include Edwards AFB and White Sands, NM for the Space Shuttle and the United States plains for Soyuz.

Ballistic Descent

Ballistic entry is a safety measure built into the design of the Soyuz. The Soyuz DM has the distinct advantage of having only one 'stable' aerodynamic position: with its heat shield down. The Apollo and Orion CMs had two stable positions, one with its heat shield down, the other with its heat shield up. If the CM were to be inverted upon descent due to a failure of the SM to separate, the crew would not survive. The Soyuz DM relied upon automated systems to maintain a shallow descent profile, minimizing the g-forces experienced by the crew. In a nominal descent, the Soyuz DM will roll to offset its center of mass and create a small lifting force, limiting the deceleration forces experienced by the crew. But if the automated systems governing re-entry and descent detect an anomaly that may make a normal entry not feasible, the spacecraft will switch to a simpler-but more stressful-trajectory, called a "ballistic" descent. In a "ballistic descent, the DM will descend through the atmosphere more quickly and impose roughly twice as many g's on the crew. The "ballistic" descent is a fail-safe mode to which the vehicle will downselect to in order to maintain the stability and integrity of the vehicle in the event of a guidance system failure upon descent. It's not a comfortable ride, but it is survivable. As a result, the steeper entry brings the spacecraft down far short of the nominal landing location and will touchdown about two minutes earlier than a nominal landing.

Figure 2.19 shows a nominal descent of the Soyuz spacecraft with a solid line along with a typical "ballistic" descent depicted by the dashed line. The controlled descent phase maintains a g-envelope of approximately 4 g's through use of controlled banking maneuvers that impart a degree of lift upon the capsule. A ballistic descent will impart approximately 8 g's of force upon the crew and will place the DM roughly 450km uprange of the intended landing site, with a downrange error of +/- 130km and a crossrange error of +/- 15km.

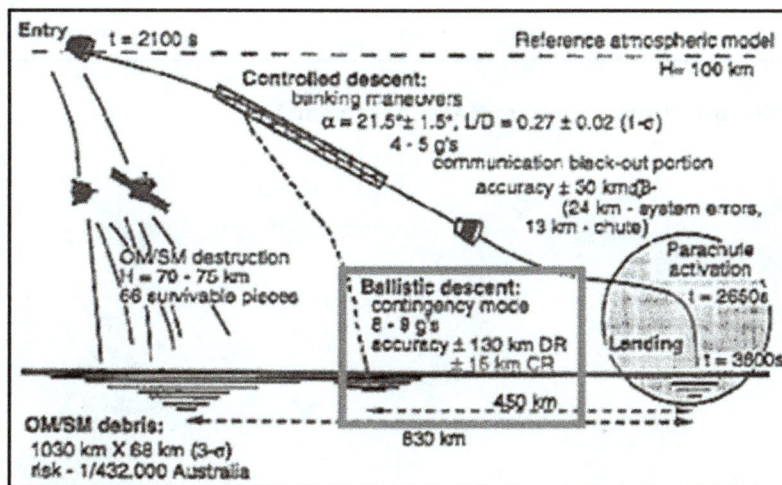

Figure 2.19 Soyuz Ballistic Entry Profiles (source: NASA documents)

Skip Re-entry

Vehicles returning to Earth from velocities in excess of escape velocity need to deplete the excess velocity. Retrograde engine firings are costly in terms of fuel used and the budgeting of this fuel in the overall vehicle mass budget, so a 'skip' re-entry allows for significant mass savings, though its proper execution relies upon a guidance and control system that could provide highly precise targeting. Skip entries also place great stress on a vehicle's thermal protection system due to rapid heating, cooling, and reheating. An undershoot would plunge the vehicle into the atmosphere at a velocity that could cause the vehicle to disintegrate while an overshoot could cause great error in the landing location or perhaps cause the vehicle to simply 'skip' off the Earth's atmosphere into deep space. The Apollo CM was capable of a skip entry, though it was never employed in a crewed mission. Later, the Orion CM was designed to be able to execute a skip entry, largely to be able to target the PLZ from greater variety of abort trajectories from lunar design reference missions.

The basic concept of a 'skip' entry is to enter the atmosphere at such an angle that the vehicle is 'pushed' back out into space, conceptually similar to a pebble skipping across the surface of a lake. Each time, the craft's velocity is reduced so that it can eventually drop into the atmosphere at a low suborbital velocity. This profile is shown conceptually in Figure 2.20. Errors in the trajectory of a skip entry can cause great error in the landing location, assuming that the spacecraft is not destroyed by an undershoot or skipped off into space via an overshoot.

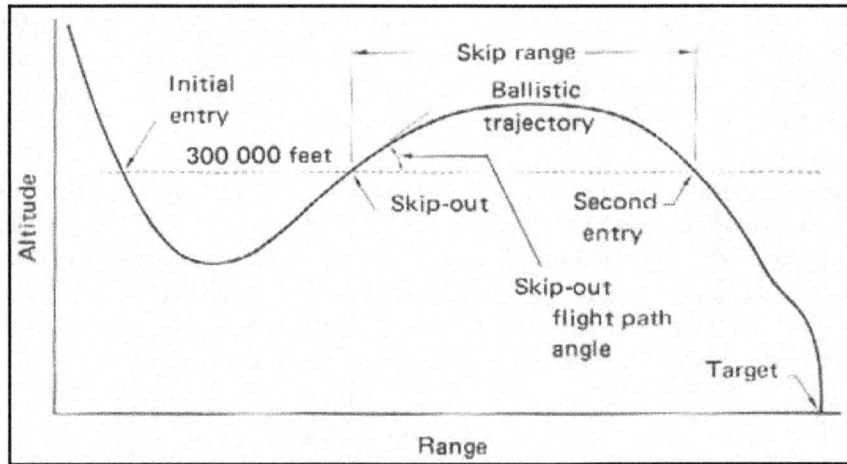

Figure 2.20 Skip Re-entry Profile (source: NASA documents)

3   Post-Landing Failure Environments

There are numerous failures that can occur during the Entry, Descent, and Landing (EDL) phase of flight, or even during the post-landing phase while the crew is awaiting recovery that could mandate an unassisted crew egress. A proper analysis of these post-landing scenarios first identifies the failure modes, or systems that can failure leading to an environment that would require the crew to egress the vehicle within a specific timeframe. Such a timeframe is determined by both the environment as well as the mitigation methods that could be used to protect the crew from the hazardous environment.

The post-landing failure contingencies that were considered for both Project Mercury and the Constellation Program are detailed here. Though both spacecraft have similar landing scenarios in that the spacecraft lands in water under the controlled descent of a parachute system, there are huge differences in the post-landing contingency planning. This discrepancy is due to the large body of experience gained through lessons learned from missions prior to the design of the Constellation architecture, as well as the use of heritage systems with known failure probabilities.

3.1   Post-Landing Contingencies Considered by Project Mercury

Project Mercury considered several post-landing contingencies but, as the egress procedures were rather unprecedented, the procedures tended to be more subjective. The principal failures considered by Project Mercury involved the failure of the parachute cutters and landing in rough seas. Other contingencies were considered, including fire and/or fumes, but there were no detailed egress procedures; the operations manual simply instructed the pilot to egress the capsule. Indeed, a one-man egress capsule would have much more simplified egress considerations that a multi-crewmember vehicle and the unprecedented nature of the flights left many egress decisions to the judgment of the pilot.

Chute Fails to Release
If the main parachute failed to release from the Mercury capsule after splashdown, the parachute could act as a sail and pull the capsule through the water, complicating rescue or even potentially sinking the capsule. The Operating Procedures detailed what the astronaut was to do in the event of a failure of the chute release mechanism after landing:

1. Reserve Deployment ring – PULL
2. If capsule is tumbling, stay in seat.
3. If capsule is not bouncing or tumbling:
    a. Remove right side of instrument panel and escape hatch
    b. Push out parachute container
4. Cut shroud lines with knife

Reserve Chute Fails to Eject
If the capsule's reserve chute failed to eject after splashdown, the astronaut was instructed to push the reserve out with the parachute container. The parachute system was easily accessible to the astronaut and could be extracted if a secondary egress path was required.

Cooling System Fails

If the cooling system failed, the astronaut was to 1) 'Remove escape hatch and open suit' and 2) 'Evacuate the capsule if it overheats'. It is notable that the suits were only designed to provide buoyancy if they do not take in water, so great care would be needed if the astronaut were to open his suit and evacuate the capsule.

Water Leaks into Capsule

Since a crewed egress from a capsule in water such as that of Project Mercury was largely unprecedented, there were numerous procedures that relied upon subjective assessments of the conditions. For example, if the capsule leaked after landing, the pilot was given these two choices:

1.   If leak is large, leave capsule immediately
2.   If leak is small, check conditions outside before deciding to leave capsule immediately.

Capsule Lands in 'Rough Seas'

If the capsule was to land in 'rough seas' the FOM instructs the pilot to "Remain in the seat with harness fastened. If forced to leave capsule, use extreme caution." Again, 'rough seas' is a subjective interpretation and the risks of attempting an emergency unassisted egress in 'rough seas' were little known.

3.2 Constellation Post-Landing Egress Environments

The Constellation Program thoroughly analyzed failures that could occur upon landing and during the time leading up to rescue of the crew and recovery of the vehicle. The component failures were grouped into groups based upon the hazardous environments that they create. These egress environments were assessed separately as to the urgency they create as well as to the specific egress procedures the crew would perform to mitigate the hazard or to perform an unassisted egress operation.

There are various failures that can occur in the post-landing phase that can be adequately mitigated without the need for egress. There were also failures that were deemed too severe to be survivable and as such were not considered. Such failures included sinking due to a rupture in the pressure vessel as it was assumed that an impact strong enough to rupture the pressure vessel, either from landing or through an external explosion of a hydrazine or oxygen tank, would be strong enough to cause a LOC event.

Though all spacecraft designed for water landings (e.g. Mercury, Gemini, Apollo) have landed in water, not the same could be said for vehicles designed to land on land. The Russian Soyuz 23, designed to land on solid ground, had the misfortune of landing in a frozen lake. Fortunately the crew was recovered alive, but only after narrowly averting a LOC event resulting from hypothermic and toxic environment conditions. The Orion spacecraft was designed to land nominally on water, much like the Apollo spacecraft had, and the possibility exists that the spacecraft could inadvertently land on solid ground, the most probable case being where winds could blow the spacecraft back to an impact on solid ground following a pad abort. In this case, a failure of one or more of the spacecraft's three parachutes would certainly cause a LOC event. A very significant probability of injury still existed for a terrestrial landing, even under all three functioning parachutes.

Failures of the response team to arrive in sufficient time to conduct a rescue operation were also considered. The contingency recovery architecture supporting the Constellation Program was based on several layers of response, integrating a NASA-directed operation headed by Air Force PJ teams that would provide a first response capability sufficient to extract the crew and secure them in RAM-Z inflatable boats. A second level of response relied upon commercial vessels organized under the US Coast Guard's AMVER program or through the SAR capability of a local foreign nation.

Individual component failures were grouped into the egress environments they create. These failures lead to egress operations initiated by 1) unmanageable water leaking into the cabin, 2) unmanageable toxic gas leaking into the cabin, 3) CO_2 toxicity, 4) crew overheating, and 5) unmanageable fire or combustion products in the cabin. Post-landing egress operations must take into account whether or not the spacecraft lands in a 'Stable One' orientation (right-side up) or in a 'Stable Two' orientation (upside down). Like Apollo, Orion was designed with an array of five righting bags that would be engaged to right the capsule, should it land in a Stable Two orientation. Nearly half of all Apollo missions landed in a Stable Two orientation upon splashdown.

3.2.1 Water leaking into cabin

Often the first danger one thinks of in a post-landing environment after a water landing is sinking. The Constellation Program deemed any impact strong enough to cause a rupture in the pressure vessel of the spacecraft to be strong enough to kill the crew. Thus, water induction into the spacecraft would either be possible through either hatch, if the crew elected to open one, or through a failure in one of several environmental systems occupying openings on the spacecraft, such as the Positive Pressure Relief Valve (PPRV), the Negative Pressure Release Valve (NPRV), the snorkel inlet/outlet, or the docking hatch valves. As all of these valves are oriented up (near the docking hatch), sinking through water induction through one of these systems would only be possible while the CM is in a Stable Two configuration. While in Stable One, the amounts that could be inducted through any of these systems was considered not significant.

If water was allowed to free-flow through the NPRV, PPRV or through the snorkel fan while in Stable Two, the crew would have to intervene as a mechanical cap was never integrated into the system. A failure of one of these systems, combined with a failure of the righting bag system, creates a very grave scenario where a Stable Two docking hatch egress may be required of the crew. If the water is relatively calm, the spacecraft could fill partly and stabilize, acting more or less and a diving bell and facilitating crew egress. A failed NPRV or snorkel fan would drive partial flooding of the CM in about one minute, after which the crew would have some time to open the docking hatch and egress. If the sea conditions are rough, however, a timely stable two docking hatch egress may be impossible.

3.2.2 Toxic gas leaking into cabin

The impact of landing, especially if one parachute is out, may be strong enough to compromise the integrity of systems holding contaminants hazardous to the health of the crew. Orion has propulsion systems that contain hydrazine for attitude control and a system that holds ammonia for environmental cooling. For such contaminants to create an egress environment, a leak of the propulsion system and a leak or passage into pressure vessel would be required. The easiest way to introduce leaking contaminants would be through the PPRV, NPRV or snorkel fans. Ammonia is only present from S+0 – S+2 hours, while it is being used to cool the flight crew. After S+2 hours, there is no ammonia left to contaminate the spacecraft pressure vessel with

Hazardous environments created by toxic gas leaks may be mitigated by venting the cabin via the snorkel fan and/or opening the docking hatch. Opening the docking hatch, however, creates a significant risk, especially in rougher seas. Mitigation methods based on ventilation are dependent upon both the quantity of the toxin and the ability to detect it. If the environment is triggered upon landing, or shortly after, the crew would have approximately 15 minutes of air that is provided through the ventilation system directly integrated to the crew members suits (the 'suit loop'). This suit loop gives the crew adequate time to assess the quality of air within the cabin before doffing helmets and gloves. It should be noted that, once the first crew member doffs their helmet or otherwise breaks the loop, all crew members will be exposed to the external environment. The crew should then act uniformly once the decision has been made to disengage from the suit loop.

In addition to air provided by the vehicle through the suit loop, each crew member has available one Emergency Breathing System (EBS) bottle that will provide up to 10 minutes of air, enough to conduct a safe egress out of either the side hatch or the docking hatch.

3.2.3   CO$_2$ Poisoning

After the ammonia is exhausted, nominal cooling must be accomplished using the snorkel. Note that a nominal rescue and recovery will be performed well before the two-hour limit as with the majority of contingency recovery scenarios, so scenarios where the crew would be still awaiting rescue after exhausting all the ammonia would be very rare (less than 1/1000). For CO$_2$ toxicity to be credible, there would have to be a contingency landing where rescue is greater than two hours away, compounded by a failure of the snorkel system or a failure precluding the activation of the snorkel system (e.g. a hazardous environment outside the vehicle or the vehicle stuck in a stable two orientation). Typically, CO$_2$ is extracted from the cabin using lithium hydroxide (LiOH). Egress operations triggered by CO$_2$ toxicity should be constrained by time.

3.2.4   Overheating

A cooling system failure will cause the crew to doff their suits, and though this may be extremely difficult to impossible for a strongly deconditioned crew member or injured crew member but it is assumed that non-deconditioned crew members could assist the deconditioned or injured ones. A power failure may lead to an overheating environment within the capsule. It was generally accepted that, in the worst case of landing in a very hot environment with a recovery operation more than two hours away, such as in the equatorial Indian Ocean, the crew may be forced to open the docking hatch to mitigate cabin temperatures should the snorkel fan system fail. However, external sea conditions may make it hazardous to stay in the capsule with the docking hatch open, so an egress operation may be advisable. On the other hand, excessive sea conditions may make the egress operation very hazardous in itself. Egress operations triggered by overheating should not be constrained by time.

3.2.5   Fire and Combustion Products

A fire could be initiated by the shock of landing or by the strains on the electrical system that will be elevated during the EDL phase of the flight. After landing and as the capsule is secured, the risk of the initiation of a fire reduces greatly as the number of potential sources diminishes as the various powered components are powered down.

The mitigation methods for fire and combustion products are similar to toxic environment unless it was a very large, rapidly burning fire. As vehicle components are being designed to limit their flammability with few components operating past fifteen minutes after splashdown, the probability of an unrestrained fire is low. As the first line of defense against a fire would be a functioning detection and suppression system, an egress environment initiated by fire would require the initiation of the fire and a failure of the detection and suppression system. Though all components have an intrinsic failure probability that can lead to the introduction of fire and smoke into the cabin, human error must also be considered as another valid failure mode.

If the crew will be unable to contain the source of the fire, then an unassisted egress must be initiated. The EBS bottles will be an essential part of an egress operation in a smoke-filled cabin as they will provide 10 minutes of air to each crewmember. Furthermore, if the crew is still integrated to the vehicles environmental system via the suit loop, some egress operations can be performed before disengaging the loop and committing to the EBS bottles.

Case Study: The Apollo-Soyuz Test Project

The Apollo capsule that was the American contribution to the ATSP was the final flight on the Apollo Space Program. The mission was considered a successful public relations opportunity for both the US and the USSR and was also executed without flaw until the three US astronauts, Thomas Stafford, Vance Brand, and Donald "Deke" Slayton, began their descent in the Apollo spacecraft.

During the ionization blackout, astronaut Vance Brand missed flipping two switches which should have deactivated the RCS which allowed Nitrogen Tetroxide gas to circulate about inside the Apollo cabin. It was discovered later that the earth landing system had failed to jettison the apex cover and drogues as scheduled. The drogues had to be ejected manually, but this was done without first disabling the RCS thrusters. As the CM oscillated upon descent, the RCS thrusters began firing rapidly to stabilize the capsule, and combustion products including a small amount of nitrogen tetroxide that were captured in the wake behind the descending capsule were allowed to enter through the cabin-pressure relief valves.

Despite the poisonous gas, the crew landed safely and Thomas Stafford extracted three oxygen masks to prevent further harm to their lungs before the crew could egress. Nitrogen Tetroxide converts to Nitric Acid when it contacts they eyes, skin, or lungs. But the spacecraft, like half of the Apollo missions prior, had landed in a 'Stable Two' orientation, meaning the crew was floating inverted in the water. Worse, the up-righting bag system had partially failed, leaving the crew suspended in this manner until the rescue team from the prime recovery ship, the U.S.S. New Orleans, could deploy from the rescue helicopter to attach the flotation collar.

As soon as the RCS system had been disabled, fresh air was once again drawn into the cabin. Brand had passed out for a few moments after splashdown, though he had quickly recovered when Stafford and Slayton placed an oxygen mask on him. The quick response of the crew to don thee emergency oxygen masks and activate the post-landing ventilation system likely prevented further damage or incapacitation by the toxic environment

Once safely on the recovery ship, the astronauts clearly showed signs of eye and lung discomfort after landing and had reported the gas caused a "burning and irritation in the eyes." The three astronauts were placed under a precautionary medical observation while on the recovery ship and then were kept under observation following the exposure for any symptoms of pulmonary endema, a lung condition that could complicate breathing.

3.3 Constellation Post-Landing Egress Planning

Once the most credible failure modes are identified, analyses are performed to optimize the resource allocation onboard the spacecraft as well as the recovery response. The Constellation Program assumed two initial conditions upon landing: that the CM was in a Stable One or a Stable Two orientation. A failure tree analysis was constructed to identify each possible failure that could initiate upon or after landing and before rescue. Each possible failure presented operational decisions that can be used to mitigate the problem, including an emergency egress operation.

The credible failure scenarios change with time after landing. Impact related events are most significant immediately after landing. Later, after the crew has had time to safe the vehicle and doff helmets and gloves as the suit loop air is exhausted, the most credible failures involve toxic interior environment conditions. After two hours (a post-landing time duration that would only occur for off-nominal landings), the cooling system will no longer maintain ideal cabin temperatures; the most credible hazardous scenarios then involve excessive heat or CO_2 toxicity.

If a crew needs to mitigate a failure environment through an egress operation, an accurate assessment of the time required to conduct the operation is needed to properly size the consumables needed and choose what type of egress may be most appropriate.

3.3.1 Failure Trees

Decision trees are often a valuable tool in the development of operational procedures when considering the possibility of multiple contingency possibilities. The two decision trees presented in Figure 3.1 and Figure 3.2 show the prominent egress possibilities that could be initiated upon landing. Figure 3.1 shows the post-landing failure tree for stable-one landings. Figure 3.2 shows the post-landing failure tree for stable-two landings. Failure environments are identified in yellow boxes while operational mitigations are depicted in grey boxes.

In addition to assisting develop post-landing mitigation methods in the vehicle, a failure tree can be used to refine operational procedures and develop training routines so that decision-making is fast and efficient. The results of the various scenarios in stable-two landings will either cause the crew to wait for the CMUS to flip the capsule into a stable-one orientation before initiating an egress or, in the event of a CMUS failure or an uncontrollable free-flow of water through the NPRV or snorkel system, will require the crew to perform a stable-two egress through the docking hatch.

3.3.2 Constellation Post-Landing Emergency Egress Scenarios

Post-landing failure modes are grouped into periods of time post-landing so that assumptions could be established concerning egress procedures and probabilities of an egress scenario arising. Three post-landing scenarios existed for the Orion capsule, corresponding to three different time periods and three different crew and spacecraft configurations.

Scenario One: Egress Initiated from Splashdown to S+15 minutes

The highest probability of an unassisted egress occurs during the first fifteen minutes following landing. Most egress scenarios in this timeframe will be driven by the shock of the landing event. Here, we assume the crew is suited with helmets and gloves on. The Life Preserver Unit (LPU) is on but not inflated, visors are down, and the crew is interfaced with the suit loop.

The Constellation Program deemed the greatest probability of an egress scenario during the first fifteen minutes following splashdown would come from the initiation of a hazardous environment from smoke and fire. If the source is not too large, it can be mitigated by extinguishing the source of fire and venting out the cabin air via the snorkel fan. Otherwise, an emergency egress would have to be performed. If Apollo is used as an analog, there is approximately a 50% chance of the vehicle landing in a Stable Two orientation, requiring the use of the righting bag system. If the smoke/fire environment is initiated upon landing, the crew may not be able to wait for the righting system to flip the spacecraft, should it land in a Stable Two orientation. The Constellation Program deemed the probability of a post-landing egress environment occurring within the first 15 minutes to be 1/532,000.

Egress operations initiated during this post-landing phase would rely upon the EBA bottles for an egress out of the side hatch and a combination of EBA bottles and suit loop air to conduct an egress operation out of the docking hatch. This combination will ensure all crew members will be protected for 25 minutes, adequate time to egress even when considering the complications of elevated sea conditions, crew deconditioning, and sea sickness. Once one crew member opens their visor, all crew members would be exposed to the cabin air and egress operations must then be completed within the 10-minutes provided by the EBA bottles. Thus, the crew should be able to test the cabin air for hydrazine and ammonia before the first crewmember opens their visor.

Scenario Two: Egress Initiated from S+15 minutes to S+2 hours

The second scenario was chosen in a timeframe starting at 15 minutes after landing, when the air provided through the suit loop is exhausted, to a time of two hours after landing, a time when the ammonia used to maintain cabin temperatures is exhausted. In this second period of time, the crew is assumed to be suited with their LPUs on but not inflated. If the air is verified to be uncontaminated, the crew may elect to remove their helmets and gloves. The fire risk is greatly reduced during this phase as many of the non- essential systems are shut down, though individual component failure probabilities increase as they are operated for longer periods of time.

The greatest risks to the crew during this phase would be a hazardous environment caused by the introduction of ammonia gas and sinking due to a free-flow of the NPRV or snorkel fan while in a stable-two orientation (in the event of a CMUS failure). Both of these failures carry a probability of occurrence of 1/478,000. The hazardous environment may be mitigated by venting out the cabin air via the snorkel fan over a 20-minute period and the sinking risk might be mitigated by manually plugging or capping the vent.

Scenario Three: Egress Initiated beyond S+2 hours

As nominal recovery operations will happen before two hours following landing, contingency landings may delay recovery operations well past two hours. In this phase, the ammonia-based cooling system will no longer operate and cooling will be performed nominally through the snorkel fan. In this phase, the crew is assumed to be suited with helmet and gloves off, if the ambient temperature is low, and unsuited if ambient temperature is high. The greatest risk to the crew during this phase is CO_2 toxicity, which would be driven by a failure of the snorkel fan system. If sea conditions permit, this hazard might be mitigated by simply opening the docking hatch, as long as no chance of the vehicle capsizing exists. Driving failures in this post-landing phase generally provide adequate time to conduct an egress operation. In the event that there is a post-landing failure of the CMUS system, the snorkel fan will be submerged and no ventilation of the cabin would be possible. Communications with the rescue team would then play a vital part in deciding if and when to conduct a stable two egress operation, which would quickly become essential in such a situation.

3.3.3 Determining the Time Required for Emergency Egress

Full scale testing is rarely able to capture all the factors that can complicate a real-world emergency egress operation. Agitated sea conditions may be simulated, but only through the use of expensive wave pools. Furthermore, very little is known by the effects of deconditioning and only expert opinions of prior flight crew that have experienced its effects may be valid. As such, tests are often conducted in 'perfect' environments with flat water, no winds, daytime lighting conditions, non-deconditioned, non-injured, and non-seasick crew, and the vehicle in a stable-one configuration with the CMUS system deployed.

Egress time requirements will then be determined through the introduction of multiplicative factors to account for these complications. The Constellation Program used a factor of three to account for sea motion, deconditioning, and sea sickness and did not factor the effects of outside surface winds, nighttime lighting conditions. Injured crew members were also not modeled. The Constellation Program concluded that no significant failures identified would create the need for a faster egress than the current 10 minutes provided by the baseline Emergency Breathing Apparatus bottle size.

POST-LANDING FAILURE ENVIRONMENTS

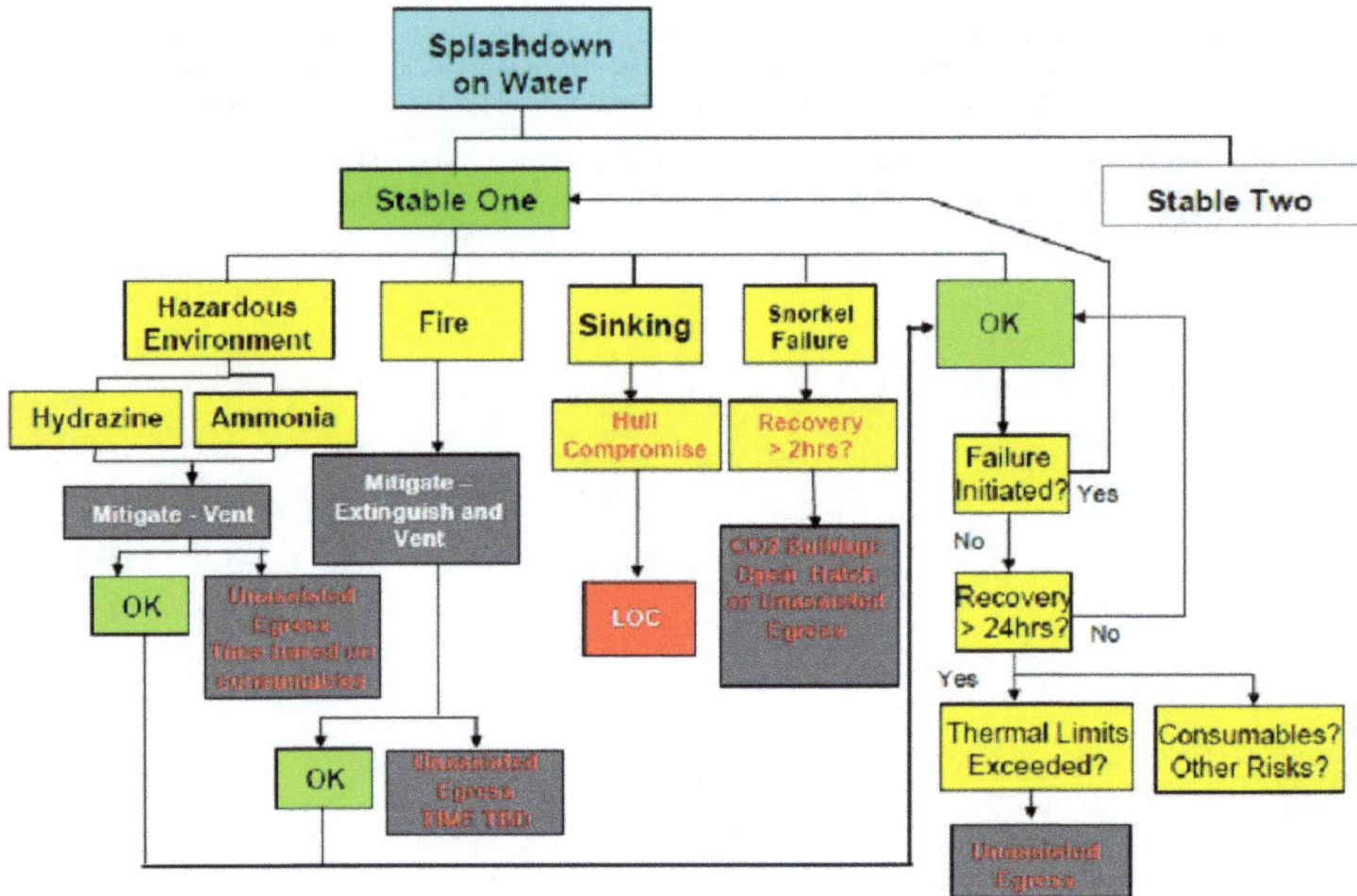

Figure 3.1 Post-landing failure tree for Stable-One Landings (Author's Models)

POST-LANDING FAILURE ENVIRONMENTS

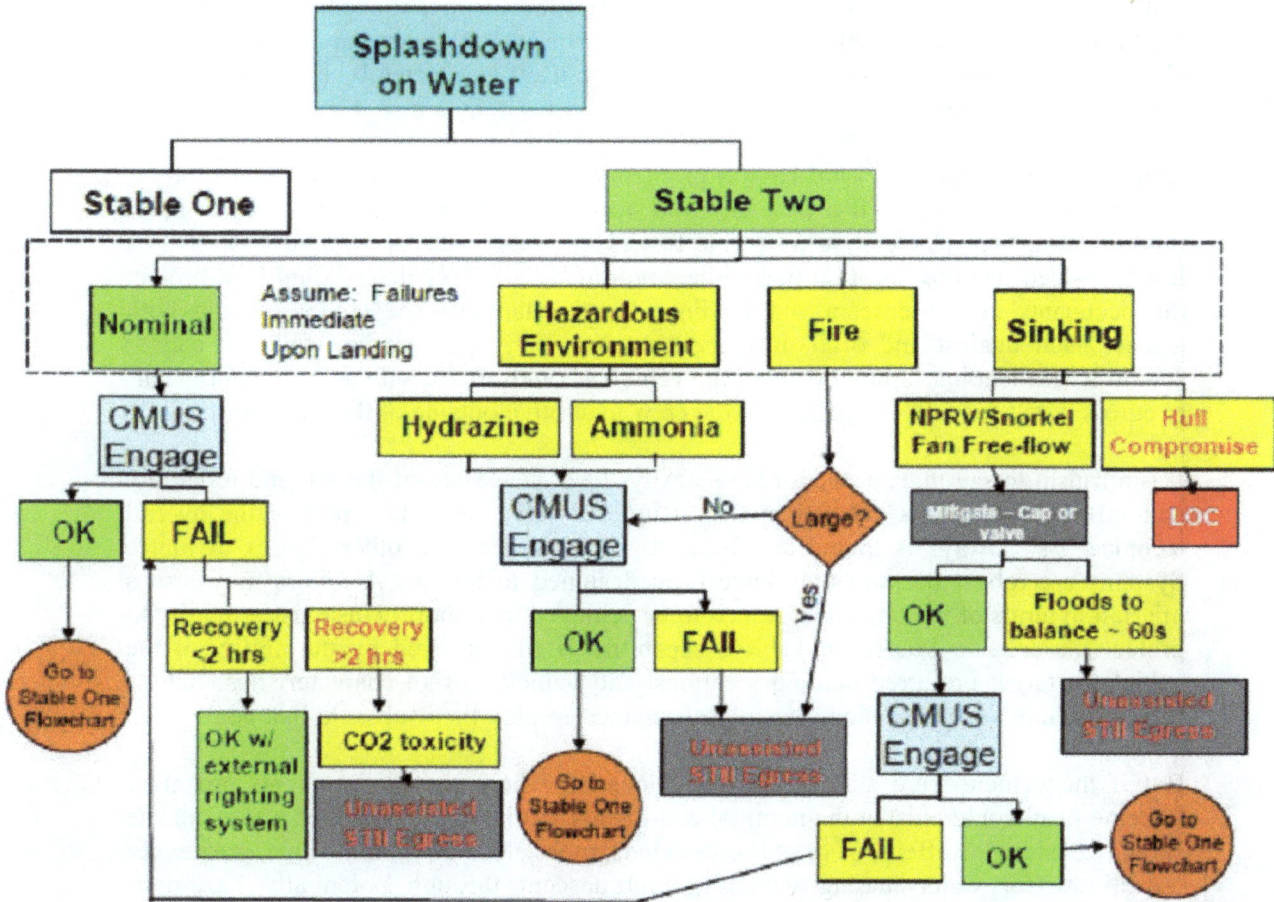

Figure 3.2 Post-Landing failure tree for Stable-Two landing (Author's Models)

69

3.4 Post-Landing Operational Considerations in Vehicle Design

Chapter One addressed the considerations and methodology used to minimize landing and post-landing risks, considering trades that affect the entire architecture, including the establishment of operational procedures and rescue and recovery architecture. This chapter addresses ways in which the vehicle design itself is affected by landing and post-landing considerations.

Ideally, a vehicle would 1) not have a problem that would cause it to deviate from its nominal mission plan, 2) if a problem were to develop causing it to assume an abort mission plan, the vehicle should be able to target an adequate landing location, 3) if a less-than-ideal landing location had to be committed to, the vehicle should best protect the occupants during descent and landing, 4) after landing, the vehicle should both protect itself against and offer mitigation means to any post-landing failures, and 5) should a post-landing failure develop, the vehicle should offer a safe and efficient means to egress with adequate equipment for the crew to survive outside of the spacecraft.

It is a truism to say that spacecraft have always been near state-of-the-art, and reliability and safety have always been of high priority, but safety as it is a part of the overall technical risk always is traded against cost, performance, and other figures of merit. Systems are robust because they have been designed to be so. A failure may lose a mission in terms of the objectives hoped to be achieved, but abort plans exist primarily to protect the crew. Post-abort vehicle capability is first defined by the ability of the vehicle to target preferred landing locations, and vehicle design characteristics such as lifting surfaces and propulsion system performance employ themselves to that goal.

But if the vehicle were still to land at a location where the environments are more extreme than would exist at the nominal end-of-mission landing zone, then a safe vehicle should consider the effects of greater crosswinds, more elevated ground slope, greater sea conditions (for water-landing vehicles), and descent through potentially hazardous environments (e.g. convective activity, icing). The seat and suit should protect their occupants through such off-nominal conditions as best as is reasonably feasible.

When considering problems that might arise after landing, hatch design and egress aid selection come into play. Egress bottles, masks, and environmental detection systems serve to mitigate many post-landing risks, but also drive vehicle mass and diminish usable cabin volume. Finally, rescue response capability can mitigate post-landing risk, but the labor and asset intensive capability can be a large driver of overall cost. Ideally, the crew would have access to equipment that could be used for survival in any random post-landing environment, but again, these come at the expense of elevated vehicle mass and decreased cabin volume.

3.4.1   Optimizing Landing Environments

The first capability that can be integrated into a spacecraft to avoid hazardous landing and post-landing scenarios is the ability to selectively target preferable landing locations. A capsule committed to re-entry and landing is simple and effective in terms of mass and cost, but offers little capability to avoid undesirable landing locations. The ability to adjust the attitude of the capsule can shorten or lengthen the re-entry trajectory and can also be used to confine the g-envelope that the crew would be subject to. One step further, the addition of lifting surfaces with control surfaces can increase downrange and cross-range capability. Finally, the propulsion system itself may be used to push the vehicle towards preferable lading locations.

3.4.2   Lifting Surfaces

Winged spacecraft reduce risk by creating an increased cross-range capability, but they also present their own unique challenges. To date, only the US Space Shuttle and the Soviet 'Buran' have flown. Lifting surfaces, however, are not as robust as capsules under the extreme temperatures of re-entry, require more complex systems to maintain the stability and control of the vehicle, and drive structural mass. As of 2012, the most mature commercial space vehicle based around a winged design is the Sierra Nevada Dream Chaser, shown through artistic rendition in Figure 3.3.

The Dream Chaser, a spacecraft originally planned in 2004 to be a suborbital vehicle modeled after the Orbital Science's X-34 and later revised to be similar to NASA's HL-20 as an orbital spacecraft, is being designed by Sierra Nevada Corporation. Dream Chaser has a 1100sm cross-range capability and is designed to be able to land on any runway at least 10,000ft in length.

Figure 3.3 Rendition of SNC's Dream Chaser docked with ISS (Credit: SNC)

3.4.3   Engine sizing

A winged spacecraft can access any landing area within the gliding range available from the altitude of the abort. A capsule-based spacecraft becomes a projectile. Though there is little a capsule can do to deviate from its projected groundtrack, energy can be added to or subtracted from the ballistic trajectory the capsule would have following an ascent abort separation. Aborts using a Launch Abort System (LAS), termed LES aborts in the Apollo Project and also referred to as 'Mode I' aborts, would place the crew at the mercy of where the LAS system propels them, though the immediate downrange regions should be benign and guarantee a safe landing. The high plains of Kazahkstan have provided this to the Baikonur Cosmodrome as the Florida coastal waters have provided to KSC.

After the LAS system is jettisoned, aborts can be performed by the SM. For the Apollo Project and Constellation Program, these were referred to as 'Mode Two' aborts. High altitude Soyuz aborts are performed in the same manner. For Apollo, a Mode II abort consists simply of a separation of the spacecraft and the launch vehicle, followed by spacecraft orientation to entry attitude and a subsequent landing in the Atlantic Ocean. In the interest of keeping a potential abort splashdown location as close to KSC as possible, a Mode III abort capability was developed that would turn the spacecraft around after separation before burning the SM engine in a retrograde direction. Since Apollo missions to the moon wanted to maintain a low inclination about the Earth, the ascent groundtracks were due east, over generally benign water, and it was almost always preferable to try to maintain a splashdown location as far uprange as possible for a high altitude abort.

The Orion spacecraft was designed first to support the ISS, which orbits at a much higher inclination – 51.6 degrees. Since the ascent groundtrack from KSC to the ISS stretches up along the Eastern Seaboard before extending across the North Atlantic, placing a crewed capsule in the rough North Atlantic is certainly something to be avoided if at all possible. For this reason, a Downrange Atlantic Exclusion Zone (DAEZ) was instituted and ascent aborts were designed to avoid this area, since a safe recovery might not be possible for a crew that might land there, especially in the winter months.

The DAEZ was defined as the area of water between the Retrograde Trans-Atlantic Landing (RTAL) zone about Saint John's, Newfoundland and the Trans-Atlantic Landing (TAL) zone about Shannon, Ireland. For a graphic depiction of these abort zones, refer to Figure 2.14. In order to minimize any possibility of aborting and landing in the DAEZ, Orion's Mode III capability was capable of both pushing forward to waters in proximity of Shannon, Ireland (the TAL zone) or turn around and perform a retrograde maneuver to a landing zone near Saint John's, Newfoundland (the RTAL zone). So if an abort trigger were to happen that would commit the CM to a ballistic trajectory that would splashdown in the DAEZ, a Mode III abort could be used. But the capability to push forward to the TAL zone or pull back to the RTAL zone would depend on the capability of the SM motor. Generally, the stronger the engine, the more the weight, cost, and complexity is introduced to the spacecraft.

The trade exists then. How powerful must the SM engine be? To answer this question fully, we should ask first what areas we definitely do not want to end up in if there were to be an ascent abort (e.g. mountains, rough or cold water, politically unstable regions, etc.). Then we should ask how much capability would be needed to assure that an abort could be targeted around the undesirable regions. Finally, if the region cannot be fully avoided, how can both the probability of entering the region and the overall risk of enter9ing the region be mitigated. The last part can be accomplished through enhancements to the spacecraft (e.g. more capable landing system, more post-landing equipment), enhancements to the rescue and recovery architecture (e.g. longer range assets, more capable assets), or through an enhanced capability to avoid the region (e.g. lifting surfaces, enhanced engine capability).

The Orion capsule had only the SM engine size to play with, since only minimal performance alterations could have been accomplished through attitude changes. Simply put, the less powerful the SM engine, the greater the possibility of entering the DAEZ and the greater areas of the TAL and RTAL zones must be protected. The greater area of these zones drives a need for recovery assets of greater range, which drives cost, and recovery assets of greater capability since the environments in a larger area have more room for variation, and waters farther offshore generally have larger swells and rougher conditions. Ultimately, this trade becomes one of overall program risk acceptance, constraints on the overall operability of the system through fewer launch opportunities, vehicle mass, and programmatic costs.

The Orion Capsule was sized with a SM engine of 7500 lbf, which was sufficient to avoid the DAEZ through Mode III targeting. For a nominal mission, the exposure to the DAEZ ranged between 22 and 28 seconds – the latest St. John's recovery occurred at 548.4 sec Mission Elapsed Time (MET) and the earliest Shannon recovery occurred at 576.4 sec MET. With a pitch-up attitude, Shannon can be reached form 570.4 seconds MET, shaving the DAEZ exposure to 22 seconds. A less powerful, 6000 lbf engine was considered, but the use of this engine would mean that there would have been a five second exposure where the CM could land in the DAEZ, and it would have been necessary to enlarge both the TAL and the RTAL zone.

The Orion Capsule was sized with a SM engine of 7500 lbf, which was sufficient to avoid the DAEZ through Mode III targeting. For a nominal mission, the exposure to the DAEZ ranged between 22 and 28 seconds – the latest St. John's recovery occurred at 548.4 sec Mission Elapsed Time (MET) and the earliest Shannon recovery occurred at 576.4 sec MET. With a pitch-up attitude, Shannon can be reached form 570.4 seconds MET, shaving the DAEZ exposure to 22 seconds. A less powerful, 6000 lbf engine was considered, but the use of this engine would mean that there would have been a five second exposure where the CM could land in the DAEZ, and it would have been necessary to enlarge both the TAL and the RTAL zone.

The Soyuz capsule had no targeting capability when the high-altitude about of Soyuz 18-1 was triggered. Consequently, the capsule landed in dangerous mountainous terrain in

Eastern Kazakhstan (see Section **Error! Reference source not found.**). Had targeting been available to the crew, a region of flatter land more suitable to landing the spacecraft could have been reached.

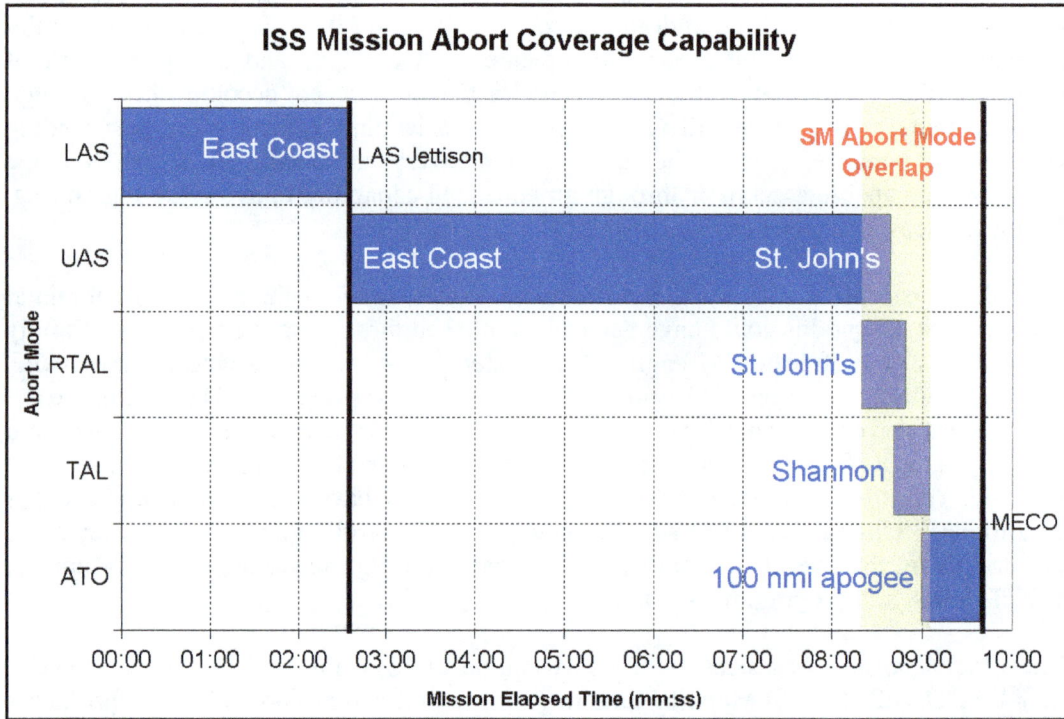

Figure 3.4 Orion Mission Abort Coverage Capability to the ISS (source: NASA document)

3.4.4   Manipulating Trajectory through Attitude

As any baseball pitcher intuitively knows, the topspin of a curveball creates a high-pressure zone on top of the ball, which deflects the ball downward in flight and causes it to drop faster. Even though the effect is mostly a visual illusion, there is a 'Magnus' force created by the spinning of the ball, exaggerated by the laces, which gives the ball more of a drop. Similarly, the attitude of a capsule can be adjusted in a way that the pressure distributions associated with the asphericity of the capsule can create deflections in its trajectory. Though the deflections may be small, they can have significant influence on a vehicle traveling at re-entry speeds. The Orion capsule was estimated to have a crossrange capability of up to 35 nm from orbital altitudes.

3.5 Minimizing Landing Risk

Seats contribute to a large percentage of the overall vehicle mass, but are essential to determining several key performance measures. Though vital for occupant protection during launch, ascent, re-entry, and landing, they are often stowed once in orbit. The Apollo Lunar Excursion Module (LEM) was designed to operate exclusively outside of Earth's atmosphere, so it deemed that seats would have been an imprudent use of mass.

In seat design, the objective of occupant protection runs contrary to the objective of rapid egress. Seats should be designed to protect the occupant under any sort of shock or impact. For launch and ascent, the conditions are generally predictable and the seat can be oriented so that the accelerative forces are oriented in a 'Gx' direction, with the force pressing the occupant back into the chair ('eyeballs in'), as opposed to the blackout-inducing 'Gz' direction, with the force oriented 'head to toe'. Winged spacecraft also have more predictable re-entry and landing characteristics, as they will only land on a suitable runway. Capsule-based spacecraft, however, can be exposed to large variations in terrain or sea conditions, especially when considering ascent aborts and re-entry contingencies. Nominal landing locations should be uniform in terrain or sea condition, as the Kazakh plains provide for Soyuz and the South Pacific had provided for Apollo. A properly designed seat for a capsule should then consider impacts along the range of terrain slopes the capsule might impact upon as well as the range of slopes along the wave fronts in the sea conditions the capsule might impact upon. The fact that the sole Soyuz 5 cosmonaut survived impact on a mountainside after an erroneous re-entry (see page **Error! Bookmark not defined.**) can be credited to the effective design of the Soyuz seats.

Because of the greater probability of landing on sloped and/or rough terrain, the seats of capsule-based spacecraft have a much greater need for the lateral support of the crew. Landing in forests or vegetation can produce abrupt lateral accelerations on impact. Sloped terrain and high sea conditions can deliver sharp lateral impacts. Lateral supports however, complicate seat design, add weight to the spacecraft, and complicate ingress and emergency egress operations. In the case of preliminary Orion tests, egress from the seats with lateral supports requires to occupant to roll to his/her side and onto knees to either crawl or stand.   Extra caution must be taken in the design of a vehicle designed to land in the water, such as the Orion capsule, as these vehicles need to safeguard against a wider range lateral G's due to the uncertainty of the slope of the water surface, which can be quite significant in elevated sea conditions.

Lateral supports also complicate the ability of a rescue crew to render assistance to the crew, which becomes especially important in the event of an injured or incapacitated crew member. Rescue crews will likely require some form of stretcher or "diving board" to assist with manipulation and intra-vehicular translation of an unconscious crewmember. Rescue crew may have severe limitations of personal intra-vehicular translation due to the small confines of vehicle, especially when complicated by the presence of lateral supports.

Lateral supports also hinder the ability of the crew to don and doff suits, umbilicals, and other interfaces. Suits are donned both while on-orbit as well as situations on the ISS where the crew might be forced to evacuate to the spacecraft and don suits once inside. Suits may need to be doffed after landing, especially if the crew were to land in high temperature conditions and in a location where the rescue might be delayed. Suits may also need to be doffed prior to an egress operation. Finally, some spacecraft provide immersion suits in the event the crew were to land in cold water and be forced to egress. If the space suits do not offer such protection , immersion suits may need to be donned inside the capsule.

Finally, lateral supports can potentially contribute to a crew member becoming trapped, as had happened in Soyuz TM-14 DM after landing, especially when considering how encumbered the crew will be with their space suits donned.

Though there is scant data available from capsule landings concerning impact direction, force, and injury potential, there ire good analog sources in auto racing as well as accident reports from auto or aircraft accidents. Though seats can be developed with mechanically removable supports, mechanisms drive weight and complexity and may not be a viable solution. Testing can be performed in laboratory environments, but considerations must be made in consideration of the actual environment in which the vehicle might land. Sea conditions, potential motion sickness, and the effects of deconditioning should all be considered.

Another trade to consider in seat design is what range of anthropometric measurements the seats should be designed to accommodate. Having seats that can support a larger range of measurements can drive seat mass and decrease workable volume within the spacecraft, but they can also accommodate a greater percentage of the populous that might serve as crew members – a broader range of scientists that might support research flights or a greater market of people that might fly as paying passengers.

3.6 Optimizing Egress Effectiveness

The ability to egress the vehicle quickly is desirable in the event of a post-landing contingency, though the design trades that allow for a quick egress often come at the expense of occupant protection during landing. Restraining devices that protect crew members during landing become obstacles after landing. Similarly, space suits that protect against decompression on entry and descent and protect the crew against hazardous environments such as fire or toxic gas become encumbrance in an egress operation. In the water, a space suit may protect against hypothermia or it can cause an astronaut to drown if the LPU is not adequately attached and inflated.

As mentioned in the previous section, a vehicle designed to land in the water, such as the Orion capsule, needs to safeguard against a wider range lateral G's due to the uncertainty of the slope of the water surface, which can be quite significant in elevated sea conditions. Complicated with the requirement to protect the crew in the event of a single parachute failure, the seats of a water-landing vehicle are often particularly confining and, as such, require more time to disengage from. So it is desirable, from a vehicle design standpoint, to create a spacecraft interior that is quickly configurable from a highly protective environment, such as that needed upon landing, to a non-confining environment, desirable in the post-landing phase. Designing systems that facilitate this transformation generally come at the expense of cost, complexity, and often most significantly weight.

Protection against post-landing failures also comes with a weight/cost/complexity penalty, so it is important to accurately identify both the most probable failures and the time that would be required to effect an egress operation under the environment created by those failures. An estimate of the time required for egress can be approximated through analog (top-down) estimates from other programs or through the use of low-fidelity mock-up tests using multipliers to estimate the effects of sea states, deconditioning, sea sickness, etc. Such tools help narrow the design trade space before costly full-analog testing is conducted.

3.6.1   Hatch Design Trades

Hatch design plays a pivotal role in determining the effectiveness of an egress operation. A hatch should be fail-safe in that it should never external environments from entering the vehicle or the atmosphere from inside the vehicle from escaping out. A hatch should also open in a way that it was not intended (e.g. Gus Grissom's MR-4 capsule). But when an emergency exists, a hatch should be able to be opened easily and rapidly from inside or from outside by a rescue team. Beyond that, a hatch should not be too massive nor bulky.

The evolution of the Apollo 1 hatch is an excellent example of vehicle design being influenced through a better understanding of the risks that might drive an egress operation. The original design of the main hatch of the Apollo spacecraft was integral to the Apollo 1 capsule and detailed in Section 0. The requirements dictated that the hatch shall be removed in less than 90 seconds. When the crew of Apollo 1 perished as a fire ravaged through the pure oxygen environment of the CM, it was clear that there were very credible scenarios where a 90-second opening time was clearly inadequate. The hatch was sent back to the drawing boards, along with virtually every other component of the Apollo spacecraft, before the crew of Apollo 7 lifted off with a totally revised unified hatch design, as detailed in Section 0.

Original Apollo 3-door Hatch Design

The Apollo 1 Spacecraft had a complicated procedure that took 90 seconds to open the hatch system, as shown in Figure 3.5. The hatch system consisted of three hatches: an inner structure (main) hatch; a middle heat shield hatch; and a lightweight outer hatch hinged to the Boost Protective Cover, which was jettisoned with the escape system shortly after launch. The inner and middle hatches had to be manually unlocked and removed to egress. Cabin pressure had to be equalized before the inner hatch could be removed. The middle hatch served as a heat shield and was an outward-opening, removable structure. The hinged outer hatch was built of fiberglass and cork and was unlocked by striking a plunger through the middle hatch that unlocked the outer hatch latches.

The Apollo 1 main (inner) hatch used internal air pressure to seal the hatch. A lip along the top and side edges and a series of latches along the bottom edge retained the hatch in the hatchway. To egress, the pressure on the inside and outside of the hatch would have to be equalized, the latches unlocked, and the hatch lifted inside the spacecraft. The middle hatch would then have to be unlocked and removed and the outer hatch would have to be unlocked and swung open. (North American Aviation engineering drawings modified by John Fongheiser)

Under good conditions the crew could unlock the doors, remove them, and egress in 60 to 90 seconds. This egress time was deemed adequate for early Apollo spacecraft as there was no requirement for extravehicular activity, and the dangers that would drive a rapid egress were not fully appreciated until after the accident. The basic procedure for egress was to:

1. Equalize pressure across the inner structure hatch.
2. Unlatch and remove the inner structure hatch and stow inside the CM.
3. Strike the plunger to open the boost protective cover hatch latches.
4. Unlatch the heat shield latches.
5. Push the heat shield hatch and boost protective hatch outboard.
6. Egress.

Revised Apollo Unified Hatch Design

After the accident the crew egress requirements were drastically changed. The crew had to be able to open the hatch in 3 seconds and egress within 30 seconds. Other requirements were dictated by schedule constraints: modifications to the existing spacecraft structure were to be minimal; no welding to the spacecraft structure would be permitted.

The selected design combined the inner and middle hatches into a unified hatch. The outer hatch, part of the Boost Protective Cover, was only slightly modified. The unified hatch mounted 15 latches linked together around the hatch perimeter. The latches applied enough force from inside the hatchway to seal the hatch. A ratchet handle allowed the crew to open or close the latches in five strokes of the handle. The handle also triggered a striker plunger to unlock the outer hatch latches (while the Boost Protective Cover was still attached).

The unified hatch design, as shown in
Figure 3.6, first flew with a crewed crew on Apollo 7. This hatch combined the inner and middle hatches and used a sturdy latching mechanism that sealed the hatch against

internal air pressure. A counterbalance improved the opening time in emergency situations. Once the latches were unlocked a cylinder pressurized with gaseous nitrogen would operate a piston to force the combined 350 pound hatch open and lock it in position. Though the new hatch greatly decreased the time it required to open in an emergency, it added 253 lbs to overall vehicle mass.

Figure 3.7 shows the Apollo 17 unified hatch interior with its latches in a locked position. The fifteen latches are linked together in four groups which can be disconnected to isolate failures. One concern engineers had of the revised hatch was the possibility that thermal exposure that could occur during a spacewalk could prevent the hatch from securing properly. To mitigate this risk, the hatch was supplemented with three small screwjacks that could secure the hatch.

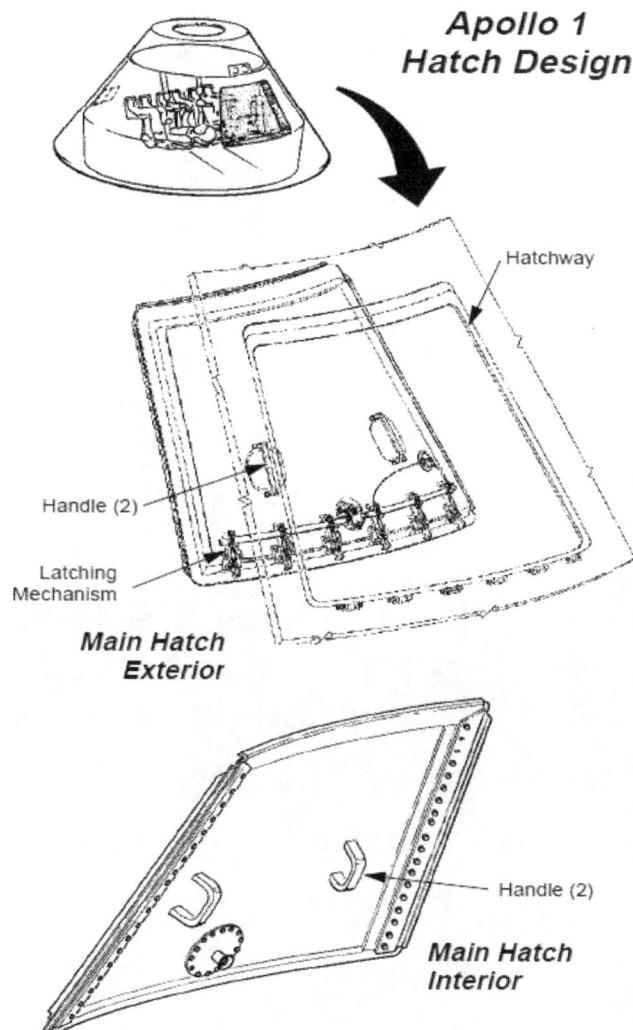

Apollo 1 Hatch Design

Hatchway

Handle (2)

Latching Mechanism

Main Hatch Exterior

Handle (2)

Main Hatch Interior

Figure 3.5 Apollo 1 3-door hatch design (North American Aviation engineering drawing modified by John Fongheiser)

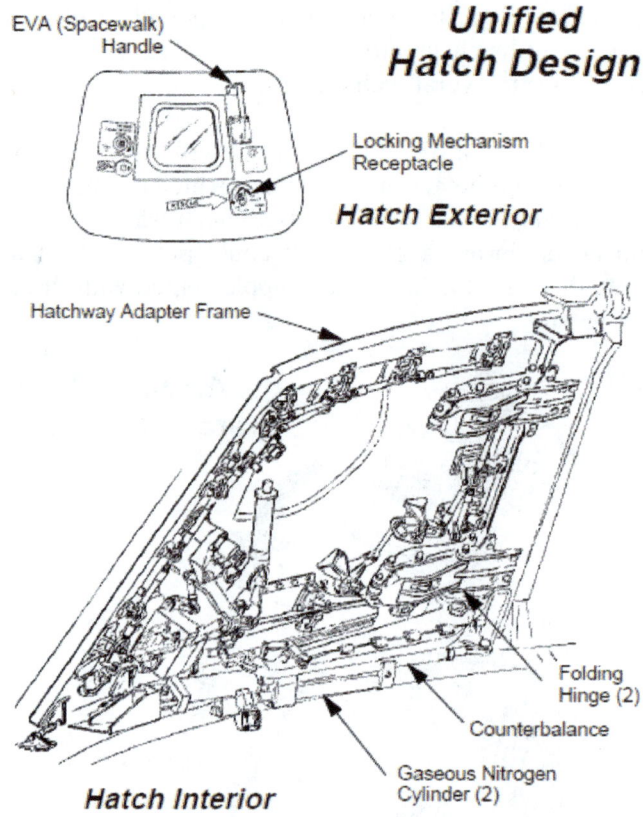

Figure 3.6 Apollo revised Unified Interior Hatch (North American Aviation engineering drawing modified by John Fongheiser)

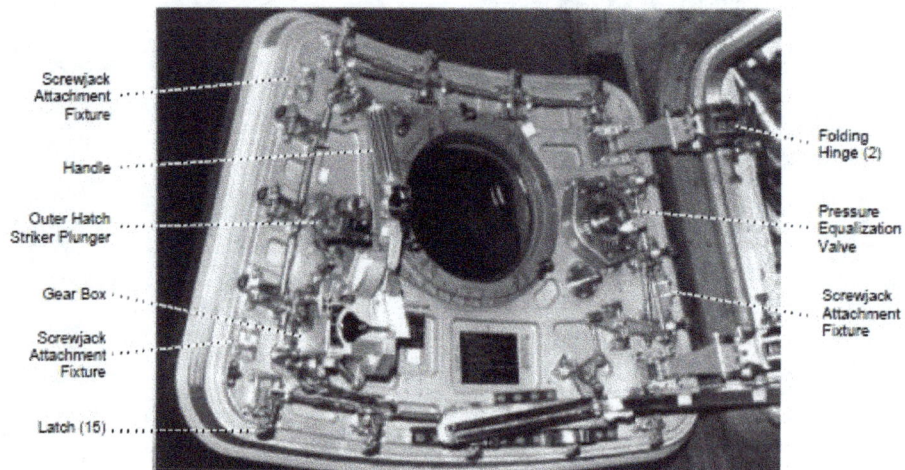

Figure 3.7 Apollo 17 unified hatch interior (credit: John Fongheiser)

3.6.2   Egress Aids

Load bearing handholds are often added to seats or to locations within the structure that might facilitate crew egress. Handholds can also be mounted to the exterior of the spacecraft to aid crew egress or the ingress of rescuers, though external handholds must be protected against the heat of reentry. As everything in spacecraft design is a trade-off, the addition of handholds adds mass to the vehicle and confines the interior volume, so straps or other lighter, more malleable handholds are generally preferred over rigid, bulky handholds.

3.6.3   Seaworthiness and stability

The stability and seaworthiness of a water-landing spacecraft also plays a major factor in the effectiveness of an egress operation. The dynamics of a spacecraft in water should be thoroughly analyzed to determine its oscillatory properties and the moment at which the center of mass of the vehicle rests above the water line, determined by the mass to displacement ratio, or the vehicles buoyancy. Though the addition of a stabilization collar and a sea anchor does stabilize the spacecraft greatly, missions to data have depended upon rescue teams to attach these recovery aids. A spacecraft system that could self-deploy such a system would greatly reduce its exposure to post-landing risk, but only at the cost of a significant weight penalty.

The hatch height relative to the water is set so that it is not too low as to present a risk of introducing water as the hatch is opened, nor is it set too high so that the egress path is complicated by greater oscillatory motion.

The spacecraft should also be stable and not prone to invert itself unpredictably. For the Apollo and Orion capsules, the uprighting bag systems served not only to right the vehicle if it were to land in a Stable Two orientation, but also to stabilize it once it had settled into a Stable One orientation so that a rough sea would not cause it to flip back into an inverted orientation.

3.7 Optimizing Post-Landing Survivability

As described in Chapter Two, there are many possible hazards that can arise in the post-landing phase, many of which have mitigation methods that assume certain equipment is available. Equipment such as Personal Egress Air Packs (PEAPs), Egress bottles, oxygen masks, and environmental detection systems all serve to mitigate hazardous environments and possibly facilitate an egress operation but all of these elements must be accounted for in the overall vehicle mass budget (Figure 3.8). PEAPs are used when the crew is suited and provide a supply of oxygen supporting a suited egress. Egress bottles or oxygen masks can be available to unsuited crew members (or crew members with their helmets removed), much as were needed by the Apollo crew of the Apollo Soyuz Test Project. Environmental detection systems are very useful in assessing the quality of the air in the cabin before disengaging from a closed suit loop system and breathing the air in the cabin.

Recovery capability does not affect vehicle mass properties but can be a very large cost driver, so it is important to make sure that investments into recovery assets produce real returns in overall risk mitigation. The effectiveness of recovery architecture can be measured by both its overall response time and the capability to conduct a rescue operation in varying environments. Investments into a greater number of primary recovery assets can improve response time, which would be useful if the vehicle is designed to support the crew post-landing for only a limited time. Investments into enhancing recovery capability using the existing assets can empower the recovery over a wider range of environments, decreasing risk and increasing the number of launch and landing opportunities.

Figure 3.8 a) Apollo 9 crew holding Personal Egress Air Packs, b) Apollo 13 Commander Jim Lovell wearing an emergency egress oxygen mask (credit: NASA)

Finally, the design team needs to decide what elements of survival gear should be included and allocated to the crew. Survival equipment trades consider the range of possible environments into which the crew may land and the equipment needed in each environment. If there is a significant possibility of lading in cold water, drysuits may be desired. Desert, tropical, cold weather, and alpine environments all require their own specialized equipment and such equipment either needs to be packed and carried or should be readily constructed from expected resources about the landing site by trained crewmembers with less massive tools. So the benefits of survival gear must consider the probability of exposure to each environment, the necessity of each piece of equipment in each environment, and the weight and cabin volume penalty of each piece of equipment.

PAD EGRESS FAILURE ENVIRONMENTS

4 Pad Egress Failure Environments

In the minutes leading to launch, configurations are rapidly made and safeguards are removed. Small triggering events can quickly lead to life threatening circumstances. Egress operations may be initiated by a failure while the spacecraft is still on the launch pad. In these scenarios, the crew has the option of egressing using an ejection seat or Launch Abort System (LAS) if the system is armed, or egressing the spacecraft out of the side hatch and descending from the launch tower via an Emergency Escape System. The Mercury spacecraft employed a solid-rocket abort tower. Ejection seats were also implemented on spacecraft supporting a limited crew size, such as were used on Vostok, Gemini, and the first four missions of the Space Shuttle. For vehicles supporting larger crews, such as Apollo, Soyuz, and Orion, LAS systems become more practical. Later flights of the Space Shuttle employed a variety of abort profiles that could be initiated.

Pad egress operations may be initiated by most of the same events as those encountered in post-landing, with the obvious exception of sinking or other water-related contingency events. The launch vehicle, however, is a huge driver of risk when the crew is in the pre-launch phase and any anomaly detected from the launch vehicle may trigger a crew abort operation, such as the abort that narrowly saved the lives of Soyuz 10-1. Further, there is usually an increased strain on the electrical system in this phase, such as that which triggered the fire that took the lives of the Apollo 1 crew.

4.1 Russian Pad Egress Systems

Russian launches have all been conducted at the Baikonur Cosmodrome. The sole cosmonaut of the Vostok spacecraft was left with only the ejection seat as a means to egress the vehicle if there was a problem in the final moments prior to launch. The access arm and support gantry could be replaced if the nature of the emergency permitted enough time, but it would have required many minutes for the structures to be replaced, the cosmonaut to egress, descend from the launch tower, and retreat to a safe distance. Seldom is the luxury of time granted to spaceflight crews dealing with a launchpad emergency, so the ejection seat would have been the only real viable means to deal with such an emergency.

The Voskhod spacecraft, however, had no means to eject from the capsule, so a timely and unlikely egress by means of descending the launch tower would have been the only means available. Fortunately, both Voskhod missions flew without need of a pad egress operation because it would have been unlikely that the crews would have survived.The Soyuz spacecraft, though a difficult spacecraft to egress from in any situation, was justly fitted with a LAS system, which undoubtedly saved the lives of the crew of Soyuz 10-1.

Case Study: Soyuz 10-1

The Soyuz launch escape system (Система Аварийного Спасения) covers the Soyuz spacecraft throughout a period spanning 15 minutes before launch to 157 seconds after launch and can be activated by either the flight crew or by the ground via a command radio-line (Командная Радиолиния) located 30km downrange of the launch site, in the event that a critical failure arises while the vehicle is on the pad or upon ascent. The main events that would trigger the system during launch are loss of control, premature booster stage separation, loss of pressure in the combustion chambers, lack of velocity and loss of thrust. Once activated, three floating struts on the payload fairing fixate to the lower structural ring of the Soyuz DM before the main escape motors fire for up to six seconds, imparting up to 21 g's of acceleration on the crew and evacuating the DM with the Orbital Module (OM) from the launch vehicle. The DM is then disconnected from the fairing, a separation motor fires and the DM falls out of the bottom of the fairing, deploys its parachute and lands in the normal manner. If the abort is triggered from the pad, the launch escape system can propel the crew to a height of 1.5km from the ground.

On September 26, 1983, the Soyuz launch escape system undeniably saved the lives of two cosmonauts, Vladimir Titov and Gennadi Strekalov, preparing to launch on what would have been the Soyuz 10 mission. This extract from *Mir Hardware Heritage* describes the only time to date a launch escape system has been used:

> *Shortly before liftoff, fuel spilled around the base of the Soyuz launch vehicle and caught fire. Launch control activated the escape system, but the control cables had already burned. The crew could not activate or control the escape system, but 20 sec later, ground control was able to activate the escape system by radio command. By this time the booster was engulfed in flames. Explosive bolts fired to separate the descent module from the service module and the upper launch shroud from the lower. Then the escape system motor fired, dragging the orbital module and descent module, encased within the upper shroud, free of the booster at 14 to 17 g's of acceleration. Acceleration lasted 5 sec. Seconds after the escape system activated, the booster exploded, destroying the launch complex (which was, incidentally, the one used to launch Sputnik 1 and Vostok 1). Four paddle-shaped stabilizers on the outside of the shroud opened. The descent module separated from the orbital module at an altitude of 650 m, and dropped free of the shroud. It discarded its heat shield, exposing the solid-fuelled land landing rockets, and deployed a fast-opening emergency parachute. Landing occurred about 4 km from the launch pad. The aborted mission is often called Soyuz T-10a in the West. This was the last failed attempt to date to reach a space station to date.*

Case Study: Soyuz 10-1 (continued)

An account from *Leaving Earth* by Robert Zimmerman:

> *It was not to be. Ninety seconds before blast-off, with Titov and Strekalov waiting at the top of their fully-fueled Soyuz rocket, a fuel valve at the base of the rocket malfunctioned, opening and spilling fuel uncontrollably onto the launchpad. A fire broke out and flames engulfed the rocket with its 180 tons of very flammable fuel. At that moment, the automatic launch-escape system should had kicked in, executing the following steps: First, explosive bolts fire, flinging the Soyuz T capsule free of the three-stage rocket. One second later, solid-fuel engines in a tower attached to the top of the capsule ignite, lifting the Soyuz T orbital module and descent module away and clear. Five seconds after that, more explosive bolts fire to separate the crewed descent module from everything else. Its parachutes then release and its retro-rockets fire, slowing the capsule enough for a safe landing.*

> *The automatic launch-escape system did not kick in, however. The fire had burned the system's wiring, preventing it from being activated automatically. Feeling strange vibrations and seeing black smoke and yellow flames outside their window, Titov and Strekalov tried to fire the launch-escape system manually, only to get no response. To fire the escape system manually from mission control required each of two different operators, located in two separate rooms, to press separate buttons at the same time. With flames rising from the launchpad and the entire rocket already leaning 20 degrees to the side, controllers scrambled madly to get the system to free.*

> *Just 10 seconds after the flames first appeared, controllers miraculously managed to somehow do this, activating the escape system and throwing Titov, Strekalov and the Soyuz T capsule more than 3000 feet into the air. For five seconds the emergency engines fired, subjecting the two men to forces exceeding 15 g's. Then the engines cut off, the descent module separated, and its parachutes unfolded.*

> *At that moment, the entire rocket and launchpad exploded. The blast was so intense that the capsule, three miles away, was thrown sideways, and launchpad workers in underground bunkers felt the pressure wave.*

Strekalov and Titov landed safely, their capsule hitting the ground with a hard bump that shook both men up but did them no damage. Rescuers quickly pulled them from the capsule, then gave them a glass of vodka to calm their nerves as everyone watched the nearby launch pad crumble in flames and smoke. It took 20 hours to put the fires out.

Figure 4.1 Launch pad abort of Soyuz 10-1

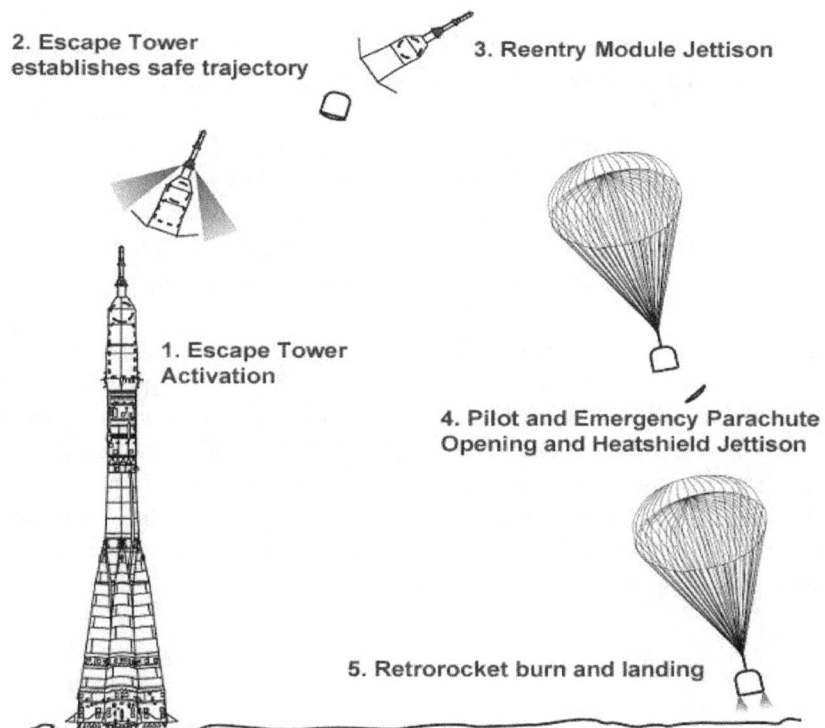

2. Escape Tower
establishes safe trajectory

3. Reentry Module Jettison

1. Escape Tower
Activation

4. Pilot and Emergency Parachute
Opening and Heatshield Jettison

5. Retrorocket burn and landing

Figure 4.2 Soyuz TMA Emergency Abort Profile

4.2 American Pad Egress Systems

Launches of Mercury and Gemini spacecraft were conducted from pads at Cape Canaveral. The Emergency Escape System (EES) on the launch pads were simple and the mitigation of launch pad emergencies was a novel field. As launch vehicles grew in both size and capability, and as the risks of the pre-launch phase became better known, more effective and responsive EES systems were needed. The Apollo I catastrophe drove a much more thorough developmental effort of EES systems that carried through the Apollo Program and subsequent crewed space programs.

4.2.1 Launch Pad Escape Systems at Cape Canaveral

The first US programs, Mercury and Gemini, lacked a rapid escape system from the launch pad; they relied solely on a LES integral to the spacecraft. For Mercury that consisted of a solid-fuel tractor motor that would pull the capsule away from the launch vehicle. For Gemini, a pair of rocket-catapult (ROCAT) ejection seats was used. But egress using a LES system has always been a risky proposition; the forces imposed upon the crew can exceed 20g's and can easily cause injury. Further, LES systems could only have been used by the flight crew after the main hatch has been secured, so they offer no protection to ground crew that might be stranded atop the gantry.

The first Mercury flights were paired with the Redstone rocket and had available a crane system called the "cherry picker", officially known as the "mobile aerial tower", which could provide an emergency escape route for the astronaut before launch. Its cab would be positioned next to the capsule hatch during the period preceding the launch, as shown in Figure 4.. If an egress was needed less than 90 minutes prior to launch, the "cherry picker" could be used to remove the astronaut. This was always a preferred means of egress over the LAS system if the nature of the emergency permitted the time. Later, for the orbital missions that used the Atlas launch vehicle, a tower ramp was used.

If time permitted, it would always be preferable for the crew to egress from the capsule and descend down the elevator in the launch tower (shown in Figure 4.). For Mercury and Gemini, such a descent took a long time – a very long time to an astronaut in an open elevator descending alongside a launch vehicle of questionable integrity. Once at the bottom of the launch tower, the astronaut would then run, fully suited, to an armored vehicle to be driven to a safe radius. It was unlikely that any true emergency would permit such time, but this was the only option available to the close-out team while they were performing final operations on the launch pad.

Figure 4.3 The "Cherry Picker" (credit: NASA)

Figure 4.4 Gantry Elevator and Armored Escape Vehicle supporting Mercury and Gemini Programs (credit: NASA)

4.2.2   Apollo Launch Pad Emergency Escape Systems

The Apollo launches were conducted at the newly established Kennedy Space Center, adjacent to Cape Canaveral. Initially, pad egress was given about the same level of consideration it was given during the Mercury and Gemini Projects. It took a major accident for NASA to investigate the need for a more capable emergency egress system at the launch pad. On January 27, 1967 a fire was triggered during a live rehearsal of the first Apollo crewed launch, Apollo 1. The fire quickly spread throughout the pure oxygen environment inside the spacecraft and claimed the lives of its three-member crew.

One of the reasons that the Apollo 1 escape hatch had no rapid jettison capability was ironically because the Command Pilot, Gus Grissom, had argued against it. Grissom's near-fatal experience on Mercury MR-4, in which electromagnetic induction caused by the contact of the helicopter support team caused the escape hatch to activate prematurely, jettisoning Grissom into the water, led him to become an advocate for the removal of explosive bolts on the hatch design of Apollo. As designed, the procedures for opening the emergency hatch required 90 seconds. The astronaut's space suits protected the crew for only eight seconds in the pure oxygen environment. Had explosive bolts existed on Apollo 1, it would have been possible the crew could have egressed the inferno in time.

In the months following the Apollo 1 accident, the Apollo/Saturn Crew Safety Panel incorporated several key modifications into the Apollo system. The escape hatch was modified to be gas-operated, and it could be opened in less than seven seconds, as opposed to the 90 seconds required by the Apollo 1 hatch. Also, two jelled water foam portable fire extinguishers and masks were supplied to the Apollo 7 crew to provide oxygen and protect from a hazardous environment. These masks were later used on the Apollo-Soyuz Test Project flight, when the crew was exposed to highly toxic Nitrogen Tetroxide upon descent.

Emergency egress procedures were improved and the launch pad was modified to accommodate. The area around the spacecraft was designed for easier access, more safety equipment was placed near the CM, and the slidewire escape system was first installed on the launch pad – a system that would carry on through the Shuttle Program and into the Constellation design. Specifically, the Crew Safety Panel proposed that 1) solutions dealing with on-pad hazards and emergencies be directed to pre-serving, as far as possible, the simultaneous availability of both a LES abort and emergency egress as escape options, and 2) time-compatibility should be a major criterion in evaluating proposed modifications to allow rapid crew egress.

The EES System that was re-designed for Apollo had to provide rapid transport of the crew and ground support personnel away from the disaster or outside the Blast Danger Area (BDA), which was defined as the area that would be affected by a catastrophic explosion of the launch vehicle. The system had to have both a quick egress time, but also have been g-limited so that further injuries might not be imposed upon the egressing personnel.

PAD EGRESS FAILURE ENVIRONMENTS

Aside from a quick egress time, the system had to:

1) Be able to handle incapacitated personnel,

2) Be passive (having either unpowered or have available back-up systems independent of launch pad systems),

3) Accommodate flight crew plus any ground support personnel and   rescue personnel present,

4) Accommodate pressurized and non-pressurized space suits, rescue suits/gear, and any special ground support personnel suits, and

5) Be single fault tolerant at a minimum, so that there could be one failure and the system still needs to operate within requirements.

The fireball expected from a Saturn booster failure was characterized by the following parameters in the Apollo Working Paper "Estimation of Fireball from Saturn Vehicles Following Failure on Launch Pad" (number 1181):
>   Nominal Fireball Diameter (ft): 1408 (Saturn V), 844 (Saturn 1B)
>   Duration of Fireball (sec): 33.9sec (Saturn V), 20.1s (Saturn 1B)
>   Peak Surface Temperature 2500 degrees (F)

Certainly, any escape system had to design around the parameters of a catastrophic failure. The early EESs were fairly simple, gravity-powered systems. Initially, a slide tube system was implemented to provide emergency egress capability to Apollo crew-members and ground support personnel that were a part of the close-out crew. The slide tube EES used the existing launch tower elevators which would transport the crew to the base of the mobile launcher platform, where they would transfer to a slide tube that terminated in an underground shock-mounted bunker below the launch pad where they could enter a sealed "blast room" to wait out the emergency.

Clearly of the same design as nuclear missile launch control centers, the Apollo "blast room" was mounted on springs and sealed with a solid steel door. The crew would strap themselves into seats to brace against a potential explosion of the launch vehicle. When deemed safe, the crew could then exit the "blast room" via a tunnel that ran to a safe distance from the launch pad.

However, the slide tube system did not handle incapacitated personnel well since there was a transfer from the elevators to a slide tube, and it did not remove personnel away from the disaster. To address these requirements, the system was soon expanded to include a slidewire system that adequately addressed both of these concerns, and it was this system that was carried into and further evolved in the Space Shuttle Program. The slidewire EES could accommodate up to nine personnel, who would ride down a steel cable to a landing site 2400 feet away.

Figure 4.5 Apollo "blast room" (credit: NASA)

Figure 4.6 Apollo Slidewire System at Launch Pad (credit: NASA)

4.2.3   Apollo Escape Options

The Apollo Emergency Ingress/Egress and Escape system were designed to evacuate the crew from the White Room on the top of the launch tower area to a protected area in the event of an impending explosion prior to the sealing of the crew inside of the spacecraft and the evacuation of the close-out crew. After sealing of the capsule, only the LAS system could be used to evacuate the crew. So pad egress decisions depended on the criticality of the emergency and the configuration of the spacecraft.

Two major categories of emergencies then exist: 1) those requiring rapid crew evacuation from the CM and 2) those requiring a rapid escape from the area of the booster. In those cases where both avenues of crew escape cannot be simultaneously present, the speed of reestablishing them is a good criterion for evaluating alternate solutions. A quick-opening hatch was designed to create an exit route within two seconds, though 35 to 50 seconds would have been needed to reactivate the access arm if it had been fully retracted. If, on the other hand, the access arm is attached to the CM, approximately 15 seconds would be required to swing the arm clear of the CM so that the LAS system could be used. Assuming the access arm was clear, a LAS abort could be performed in under two seconds.

If the situation was not critical, the crew could remain in the spacecraft until the difficulty is resolved.   If the situation was critical and the access arm was stowed, the crew could use the launch escape system to affect a pad abort. If the access arm was still in place, the crew could have the ground personnel team return to the launch pad and assist with astronaut egress via the CM Access Arm, time permitting, or the crew could perform an unaided emergency egress via the Access Arm if the situation was deemed time critical or potentially hazardous to the ground personnel team.

If the crew were to egress the CM, two options were available and the selection would depend on the criticality of the emergency. The first option was to descend more than 300 feet by elevator, a procedure that took only 30 seconds but could have felt as an eternity near an unstable launch vehicle. At the base of the launch tower, the crew would enter protected recovery vehicles could transport the crew to a safe radius, or they could rely upon the use of escape tubes that led from the base of the tower to a blast room mounted on coiled springs, much like nuclear missile control rooms, and that could absorb the explosive force of the Saturn V rocket above. A second egress option made use of a slidewire system, where three slidewire baskets would carry each of the crewmembers away to an armored personnel vehicle which could transport the crew to a bunker 1.5 miles away.

Choosing best escape path

In considering what would be the most suitable egress method to avert an on-pad emergency, the configuration of the CM and the access arm (AA) must be considered, as with the location of the close-out crew. Obviously, the LES system could not be used if the closeout crew were still in proximity. Table 4.1 shows the delays associated with initiating an egress procedure. These times represent only the time required to initiate the procedure, no the total time of egress. The actual time to conduct an egress is determined through testing, though an LES abort clears the crew of the launch vehicle in less than a second. The time associated with a CM evacuation procedure depends on the nature of the ambient environment and the condition of the crew.

On-pad configurations have been specified in terms of AA position, Closeout Crew gross location, and CM hatch position. For each configuration, the delays have been shown both for initiating LES abort and for CM evacuation and in each case, the pacing item is included. When the hatch is closed, abort must be delayed by a time ('V' 20 - 60 seconds) to get the Closeout Crew to an area offering protection from the LES rocket plume. Once the AA is clear of the CM, the only delay to LES initiation would be flight crew reaction time. If the AA has not been retracted, a LES abort would be delayed by 15 to 12 seconds in order to retract the AA, since the AA would constrain the CM, creating a potentially catastrophic event.

For a CM evacuation, a reaction time of one second is assumed and, if the hatch is closed, 2-3 seconds would be required to establish the egress route. If the AA had been retracted from the CM, its reattachment drives the egress time, as roughly 20-30 seconds are needed for reattachment of the AA. If the AA is held just clear of the CM, as in case two, 20 to 30 seconds would be required to establish an emergency egress route, which would have been insufficient to respond to an emergency like the one of Apollo 1.

Access Arm	Latched to CM			Unlatched, Clear of CM Gap Bridged No Bridge		Unlatched, Fully Retracted
Closeout Crew	Latched to CM		OFF UT			
CM Hatch	OPEN	CLOSED				
Time to Initiate LES Abort	NOT AVAILABLE	~ 20-30 sec Closeout Crew to Protection from LES Plume	~ 15 sec AA to Unlatch and Clear CM	~ 1 sec Crew Reaction Time	~ 2-3 sec Crew Reaction Time	~ 2-3 sec Crew Reaction Time
Time to Initiate CM Evacuation	~1 sec Crew Reaction Time	~ 2-3 sec Hatch Opening	~ 2-3 sec Hatch Opening	~ 2-3 sec Hatch Opening	~ 20-30 sec AA Close Gap and Latch	~35-50 sec AA Traverse to CM & Latch

Table 4.1 Relationship between selected On-Pad Configurations and Time Delays to Initiate Flight Crew Escape

The importance of an emergency egress system was brought home to the American public in 1967 during a preflight test of the Apollo 204 CM, which was intended be flown as Apollo 1. With astronauts Virgil Grissom, Edward White, and Roger Chaffee onboard, this would have been the first crewed Apollo mission, but on January 27, 1967, during a preflight test on the launch pad, a fire swept through the CM killing all three astronauts.

Earlier that day the crew entered CM, perched atop its massive Saturn V rocket, to conduct a final series of tests. Since Grissom's spacesuit oxygen loop had a 'sour smell', the crew stopped to take a sample of the suit loop, and after discussion with Grissom decided to continue the test. Then a high oxygen flow indication periodically triggered the master alarm. The matter was discussed with environmental control system personnel, who believed the high flow resulted from movement of the crew. The matter was not really resolved. Faulty communications continued between Grissom and the control room for most of the test, even though the crew made adjustments. Later, the difficulty included communications between the operations and checkout building and the blockhouse at complex 34, which forced a hold of the countdown at 5:40 p.m. At 6:31 the count was about to restart when ground instruments showed an unexplained rise in the oxygen flow into the spacesuits.

Four seconds later, an astronaut announced almost casually over the intercom: "Fire, I smell fire." Two seconds later, White's voice was more insistent: "Fire in the cockpit." White instantly began to unlatch the side hatch, which involved moving the headrest out of the way, operating the ratchet-like equipment to open the latch, and removing the latch. The CM ruptured. Flame and thick black clouds of smoke billowed out, filling the launch escape system atop of the spacecraft. The rescue teams had difficulty getting close to the module because of the intense heat and dense smoke but when the hatch was finally opened, all three astronauts were dead. The primary cause of death was carbon monoxide poisoning, with thermal burns having contributing effects.

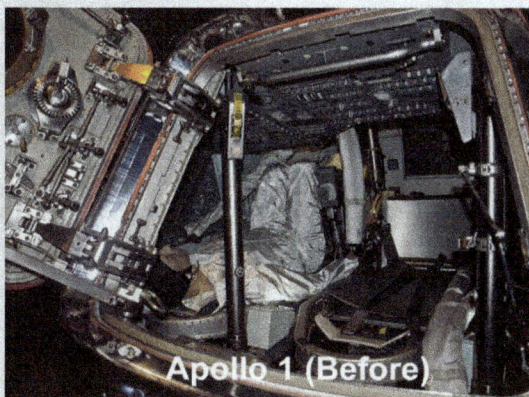

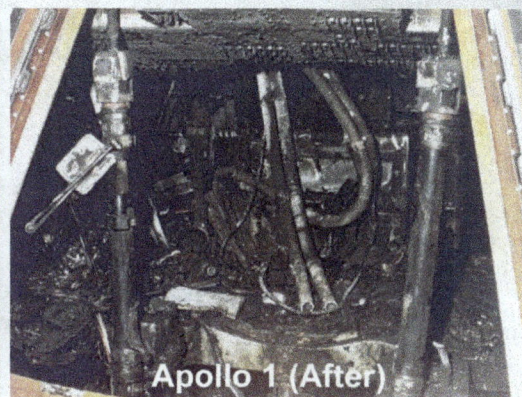

Apollo 1 (Before)   Apollo 1 (After)

Apollo 1 Aftermath

After removing the bodies, NASA impounded everything at launch complex 34 and on 3 February, NASA Administrator James Webb set up a review board. Engineers at the Manned Spacecraft Center duplicated conditions of the capsule as it was before the accident, without of course the crewmen inside. The reconstructed events and the investigation that followed showed that the fire started in or near one of the wire bundles to the left and just in front of Grissom's seat on the left side of the cabin. The fire was probably invisible for about five or six seconds until Chaffee sounded the alarm.

Apollo 1 Contributing Factors and Recommendations

As with most engineering disasters, the determination of the exact cause of the disaster was not easy, and the investigation delayed any future launches until the CM could be re-certified for flight. Even to this day, the specific initiator of the fire has yet to be determined, and it probably never will be. However, after investigation, a number of factors that significantly contributed to the disaster were determined. The fire began because of an electrical short and it spread quickly because of the volatile atmospheric conditions and the presence of combustible materials inside the module. The hatch design and NASA's management of the Apollo program may have also been contributing factors. From the investigation the review board came up with recommendations that would prevent future disasters. The following are some of the main recommendations that were put forth by the review board.

1. Oxygen or any other material that is combustible should be restricted and controlled.

2. That combustible materials used should be replaced wherever possible with non-flammable materials.

3. That atmospheric conditions of 100% oxygen are not permissible on tests.

4. Spacecraft should be designed with safety as a primary consideration.

5. In future the time to escape be decreased through other escape routes and a hatch that can easily open under extreme pressure difference.

6. It was recommended that all personnel should have proper training and should practice for emergency egress procedures.

It remains to be seen just how much emphasis will be place on emergency egress training when it comes to training future commercial astronauts. Much will depend on what egress capability exists, as some vehicles will no doubt be easier to escape from than others. For example, the Space Shuttle, which, until 2011, routinely ferried astronauts into space, had an extremely limited emergency egress capability whereas the Orion capsule had two viable escape options and an enhanced Emergency Escape System at the pad.

4.2.5   Space Shuttle Pad Emergency Egress System

As there was no launch escape system available to the Space Shuttle crews as there had been for Apollo crews, any emergency on the pad would require the crews to escape using a slidewire system that was enhanced from the system used by Apollo. The Shuttle EES consisted of five slidewires (and later expanded it to seven) with baskets that could hold up to 4 people each. These slidewires ended at the same Apollo bunkers outside the BDA where personnel could wait out the disaster evacuate further via an M-113 armored vehicle.

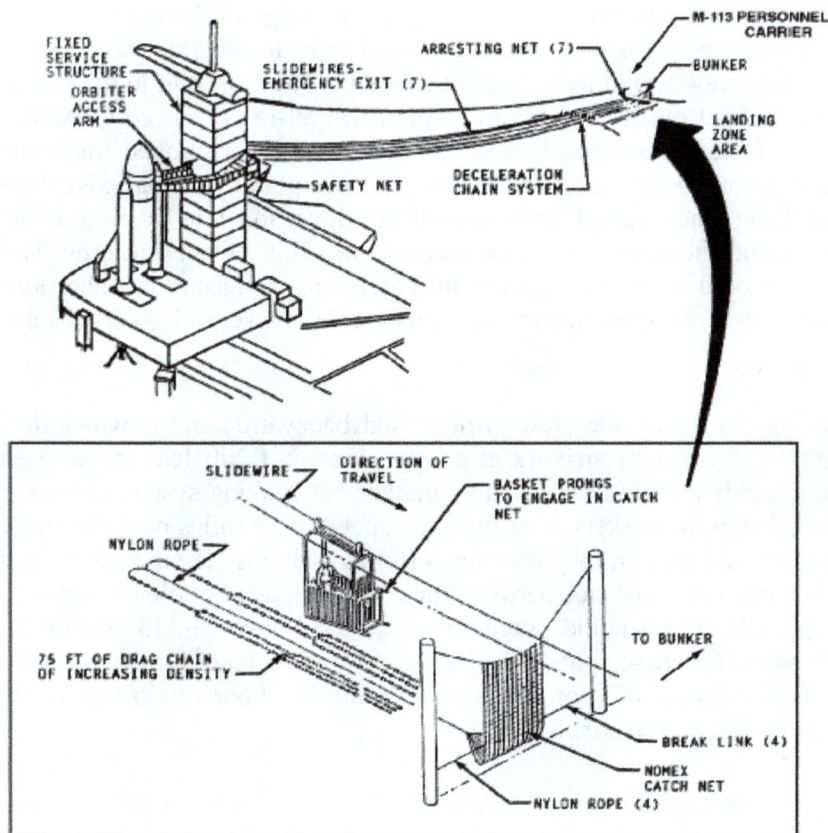

Figure 4.8 Space Shuttle slidewire system (credit: United Space Alliance)

97

Figure 4.9 Shuttle EES Baskets (credit: NASA)

For Space Shuttle crews to perform an emergency pad egress, the crew access arm would first need to be in position, but this is removed only in the final seconds of a launch countdown. The crew would then detach themselves from the suit loop system, open the egress hatch, and get out of the Orbiter in their full 80-lb Advanced Crew Escape Suits (ACES). This is not an easy task since the crews will be positioned 'on their backs' as the Orbiter is mounted vertically before launch. Temporary netting exists (see Ch 5) to facilitate an emergency egress from launch configuration that is placed into position before launch and then removed once securely in orbit. Once the crew is out of the Orbiter, they would cross the gantry in potentially a hazardous and low-visibility environment to the EES system, and then enter one of the seven baskets (backwards) and release it.

Upon releasing the basket the crew is propelled backwards and downwards from their initial height of 195ft until arriving at a large sandpit 1,200 feet to the west near the entrance to a highly armored and sealable bunker. A braking system catch net and drag chain slow and stop the baskets from their top speed of 55 miles per hour in about half a minute. Once in the bunker, the crew may elect to wait out the emergency in the bunker or opt to put some more distance between them and a Space Shuttle containing 1,100,000 pounds of propellant that could potentially explode. Two M-113 Armored Personnel Carriers, shown in Figure 4., exit on the back side of the bunker for this purpose, and they can each travel at speeds of approximately 40 miles per hour to a triage site where they could be met by rescue personnel.

4.2.6   Space Shuttle Redundant Set Launch Sequencer (RSLS) Aborts

The Space Shuttle's three main engines are ignited prior to the ignition of the two Solid Rocket Boosters (SRBs), which irrevocably commit the vehicle to launch. From the ignition of the three main engines 6.6 seconds prior to launch to the ignition of the SRBs, onboard computers perform a final state-of-health check, much like a pilot performing a run-up before taking off. If any parameter of any of the three main engines is not within nominal limits, the computer performs a "Redundant Set Launch Sequencer Abort", where the thrust is terminated and the flight systems are automatically put in a 'safe' mode. Had the SRBs ignited otherwise, the vehicle would be committed to launch in a potentially underperforming state, leading to a hazardous RTLS, TAL or even a contingency abort initiating no earlier than 123 seconds after launch. RSLS aborts have happened five times:

1. STS-41-D, (June 26, 1984), caused by a sluggish valve in main engine No. 3.

2. STS-51-F, (July 12, 1985), caused a problem with a coolant valve on main engine No. 2.

3. STS-55, (March 22, 1993), caused by the detection of a problem with purge pressure readings in the oxidizer pre-burner on main engine No. 2.

4. STS-51, (Aug. 12, 1993), caused by the failure of one of four sensors in main engine No. 2 which monitor the flow of hydrogen fuel to the engine.

5. STS-68, (August 18, 1994), caused by the detection of higher than acceptable readings in one channel of a sensor monitoring the discharge temperature of the high-pressure oxidizer turbopump in main engine 3.

Nothing is as deafening to a crew braced for launch on top of over 1,200,000kg of propellant than the silence that permeates the orbiter after an RSLS abort, but the crew is not clear of danger. Thrust termination and automated safing of the Shuttle's three main engines is a complicated procedure involving huge quantities of volatile propellants. Egress of the crew must be performed in a safe and timely manner.

Figure 4.10 Space Shuttle Emergency Escape System (credit: NASA)

Figure 4.11 Space Shuttle M-113 Armored Personnel Carrier (credit: NASA)

4.2.7 Constellation Program Emergency Escape System

NASA's Constellation Program designed a highly advanced EES, though the program was cancelled before its construction. The Constellation Program requirements call for astronauts to be able to get out of the spacecraft and into the bunker within 4 minutes. The Constellation Emergency Egress System is fundamentally a group of multi-passenger cars on a set of rails reminiscent of a modern roller coaster standing roughly 380 feet above the ground. It consisted of a multi-car gravity-propelled high speed rail system to evacuate the crew to a safe haven, as shown in yellow in Figure 4.2. The system imposed only moderate g-forces and used a passive electromagnetic braking system to transport the crew to a triage site outside the BDA. One of the benefits of the rail system is that the track can take the astronauts directly to the bunker door – a huge benefit if one of the crew members or a ground crew member was incapacitated.

In addition to providing a quicker, yet g-limited egress, the Constellation EES incorporated several design improvements over the prior Shuttle EES. Specifically, the Constellation EES:

1) Eliminated any steps or changes in elevation for walking personnel until they enter the escape car.

2) Provided rescue and egress aides in close proximity that could be quickly put in place to help extract crew from the vehicle, or to render assistance to an injured or incapacitated crew-member.

3) Provided enough room on access arms or along the egress path so that there would be no congestion of egressing crew-members and rescue personnel and enough room to accommodate equipment that could be required to assist an injured or incapacitated crewmember.

4) Kept the egress path shielded from heat, fire, and debris that could be ejected from the vehicle and provide a water deluge system (with a non-slip surface along the egress path) to mitigate such risks.

Figure 4.2 Constellation Program Rail EES (credit: NASA)

5 Egress Systems

In this chapter, a survey of the evolution of egress systems is provided, starting from the simple designs of the Vostok and Mercury spacecraft and leading to the more complicated systems integral to the Space Shuttle. Unique design elements that comprise relevant components (e.g. seats, seat interfaces, hatches, egress aids, landing system elements, recovery system components, and stabilization components) of each spacecraft are presented in this section. Components described elsewhere, or components that have had little design evolution from prior vehicles, are not described.

5.1 Vostok and Voskhod Egress System

The Vostok spacecraft successfully carried the first six Russian cosmonauts into orbit, starting with Yuri Gagarin on 2 April 1961. The Vostok spacecraft had no provisions for a safe landing, and though this simplified greatly the task of vehicle recovery, it did require the cosmonaut to rely on the same system that was integrated to the spacecraft to conduct a launch abort: an ejection seat. A launch tower was considered in the design phases of Vostok, but was later discarded in an effort to reduce the overall mass of the spacecraft. Forty seconds after launch the vehicle was traveling fast enough and high enough to render the ejection seat ineffective; there was no safe way to egress the spacecraft until it had achieved orbit, re-entered, and descended to a sufficiently low altitude. At this point, the cosmonaut was required to eject and make a separate parachute landing (Figure 5.1). A key feature of the Vostok seat system was that it was oriented in a semi-reclined position, 65° to the horizontal. This acted to minimize the acceleration loads on the cosmonaut during launch, as well as ensured the correct orientation during ejection from the DM.

In 1964, the one-man Vostok spacecraft was adapted to accommodate multiple crew members. The new variant was named Voskhod, but with multiple crew members on board, there was no way to provide ejection seats to each crewmember. The rapidly advancing American Space Program was creating political pressures within the Soviet Union, the same pressures that later contributed to the death of Vladimir Komarov on Soyuz 1, led to a Voskhod spacecraft that had no provisions for emergency escape or recovery from launch vehicle failures. In the interest of flying multiple crewmembers to keep pace with the Americans, Voskhod was built with no mass margin for ejection seats and space suits and their crews flew into space without anything more than street clothes. To date, the two Voskhod crews have been the only space crews to fly into orbit without such protections [4].

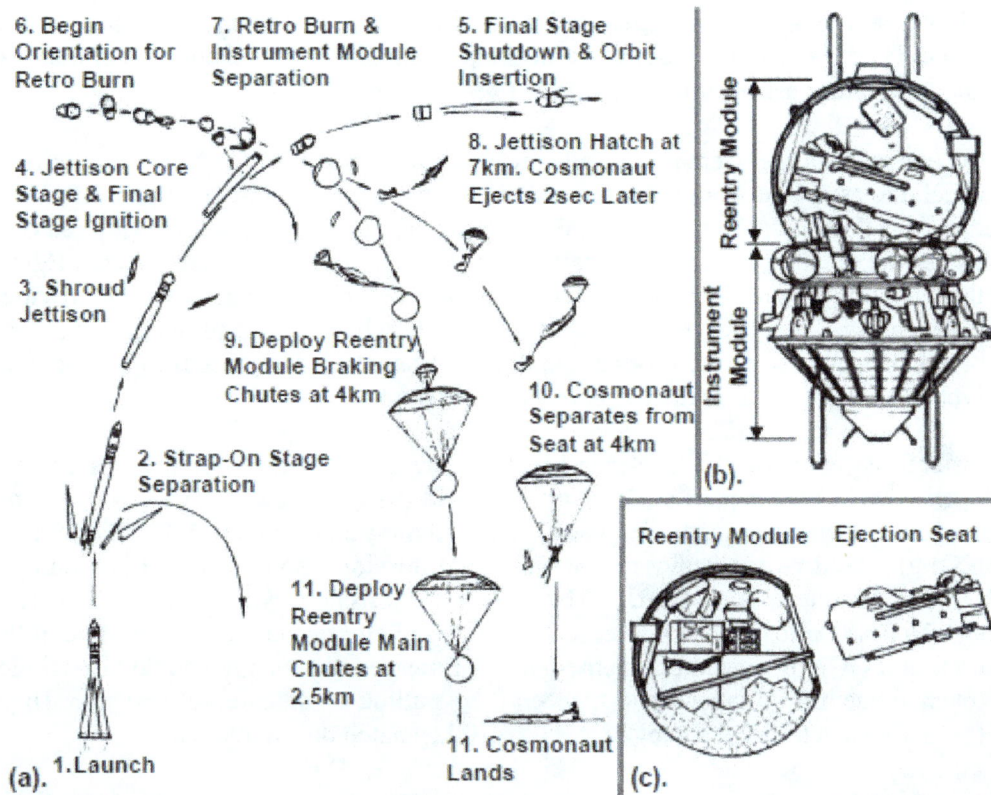

Figure 5.1 Vostok Mission (a). Mission Profile (source: Hall, R., and Shayler, D., 2001) (b). Detail of Vostok Vehicle (source: Siddiqi, A.A., 2000) (c). Detail of Seat Ejection (source: Siddiqi, A.A., 2000)

5.2 Project Mercury Egress Systems

The Mercury capsule supported one pilot through a total of six crewed flights; two suborbital flights used the Redstone launch vehicle and four orbital flights used the Atlas launch vehicle. Project Mercury performed extensive research to develop a lightweight system which could minimize the level of sustained acceleration experienced by the astronaut during launch and reentry. Not having prior experience with crewed rocket flights, the Mercury Project decided to build the launch escape system to a performance that could be survived – barely. A series of centrifuge tests were first conducted to quantify the limits of acceleration tolerance in humans, which revealed that humans can tolerate great accelerations best if they are positioned so that the acceleration loads would be applied through the chest (+Gx). Results of these tests led to the final Mercury molded fiberglass couch design that positioned the occupant in a semi-supine position.

As the accuracy of re-entry targeting and landing was less known, the astronaut was well prepared to survive within a wide range of potential environments. Aside from the ability to survive for long periods of time in water, the astronaut was trained to survive in

103

tropical or desert areas for prolonged periods and to aid rescue operations using various signaling devices. A detailed description of the post-landing survival equipment that was available to Mercury astronauts is provided in Section 5.6.1.

Initially, egress from the Mercury capsule required the removal of the forward bulkhead, after which the astronaut would climb through the antenna compartment into a raft. This procedure was deemed difficult for a healthy astronaut, but precarious for an injured one. For this reason, a 70-bolt, pyrotechnically activated side hatch was developed. Egress from the Mercury capsule was then possible both through the side hatch, which provided a quicker egress, and also though the egress path where the instrument panel and parachute bays could be pushed forward and the astronaut could egress through the apex of the vehicle, a slower but arguably more stable method.

Two different systems of quick-release hatches were used for the Mercury spacecraft's side hatch. On Alan Shepard's MR-3 mission and prior uncrewed missions, a simple design with a mechanical latch was used. The latching hatch of the MR-3 capsule, at 69lbs (31kg) would be too heavy to use on orbital missions, so a new design would be fitted into subsequent spacecraft. The following mission, Gus Grissom's MR-4, introduced a new explosive hatch release, that allowing an astronaut the ability to exit the spacecraft quickly in the event of an emergency. Emergency personnel could also trigger the explosive hatch from outside the spacecraft by pulling on an external lanyard. There were thus two ways to fire the explosive 23 lb (10 kg) hatch during recovery.

A post-landing emergency egress was conducted on the second crewed Mercury flight, Gus Grissom's suborbital MR-4 flight almost ended in disaster after the astronaut was forced into the water when the explosive bolts that could eject the side hatch suddenly activated when the rescue team first made contact with the floating capsule. Grissom escaped the rapidly sinking capsule but found that, after several minutes in the water, his suit was progressively taking on water. In a near-panic to keep his head above water, he was finally hoisted aboard the rescue helicopter.

Gus Grissom's MR-4 "Liberty Bell 7" capsule was the first Mercury capsule to be equipped with an explosive hatch, since Alan Shepard's MR-3 capsule was not so equipped. The hatch could be activated, but only after several very intentional actions. First, the hatch would be armed by removing the detonator cover and removing the protective pin from the detonator plunger. The hatch could then be jettisoned. From an interview given by Wayne Koons, the Flight Range Coordinator under the Flight Operations Division at the time of the flight:

> *On that hatch there was a safety pin. It's kind of like the safety pin on a hand grenade. You know, you slip it out to enable the control to move. There was also, I think, a cover over the button that you used to explode the hatch. The hatch had this initiator in there, but what actually separated the hatch was a ring of Primacord that went all the way around the hatch. Then it was secured with a bunch of bolts that had been weakened by being [necked] down. The Primacord*

caused the bolts to fail, and so the hatch was really bolted in place and sealed. Then when you exploded the primer, which set off the Primacord, it blew the hatch off. There'd been some tests done, and the thing came off, a big bang, and it really blew the hatch a good distance.

So anyway, he had taken the cover off and pulled the safety pin preparatory to getting out. But there's still, the way that thing's configured, there's a ring around that button, and it just doesn't seem possible that he made contact with that. Certainly, based on his debrief, he certainly didn't do it deliberately, and it's hard to see how he could have possibly done it accidentally, because the way that button is down inside a ring, a cylindrical protector, you have to be very deliberate about it to make that thing go off. (Note: The detonator cover and survival knife were both found inside the recovered capsule.)

5.3 Project Gemini Egress Systems

The two-man Gemini spacecraft initially was to use a launch tower to perform pad or ascent aborts, though this design was later dropped in favor of an ejector seat design which would provide a means for crews to egress the capsule from pad aborts at zero velocity to heights in excess 15,000ft, where the spacecraft would be traveling at about 0.75 Mach. The window when these 'Mode I' aborts could be performed lasted 50 seconds after launch. Afterwards, ejection would not be an option as the vehicle would be traveling in transonic and supersonic speeds, so a second means of abort was used that involved upper stage ignition that would push the spacecraft away from the launch vehicle after the launch vehicle shut down. This was termed the 'Mode II' abort. 'Mode III' aborts involved a retrograde burn after separation to bring the splashdown point closer to benign waters and recovery forces.

5.3.1 Gemini Ejection Seats

The ejection seats used in the Gemini spacecraft were designed to provide restraint to the crew during nominal operations and also provide a means of escape during ascent and landing. The decision to abort was entrusted to the crew, which simplified the design but raised uncertainty of whether or not the crew could respond quickly enough to initiate an abort. Once in the upper atmosphere the seat firing handle, which was located between the astronauts' legs, was stowed so it wouldn't be inadvertently actuated in space. Once initiated, a hatch actuator system would open the egress hatch above each crewmember and, through use of detonating fuses, would drive control rods that would lock open the hatches. This system was validated up to altitudes of 70,000ft (21340m).

The ejection seats were designed using calculations of the maximum fireball that could be expected in the case of a launch pad explosion, which was deemed the most critical abort scenario. The result was the most powerful ROCAT ever developed in the United States for propulsion, a seat that could outrun the fireball for a distance of about 800 feet - fast enough that the nylon parachute would not be damaged by the heat pulse of the exploding fuel. The system was tested extensively by technicians using boilerplate capsule mock-ups which were launched at high speeds down sled tracks and seat ejections were filmed to verify high speed ejections were possible. Boilerplate capsules were also mounted in launch attitude on top of a high tower and seats were fired across the desert to prove the 800 foot mark could be reached, and that the parachute system would deploy properly. Also included in the ejection seat structure was an egress kit and survival equipment. The egress kit actuated upon activation of the ROCAT and provided the astronaut with oxygen and pressurization during aborts at altitude. The survival kit provided essentials for the crew to survive post-landing while awaiting rescue.

5.3.2   Gemini Landing and Recovery Systems

An inflatable Rogallo Wing concept was investigated in the early stages of Gemini's development in hopes that the spacecraft would be able to glide down to a terrestrial landing, but this concept had to be discarded in 1963 in favor of a parachute system similar to that used during Project Mercury. Project Gemini used a three-parachute system where the main parachute suspension is shifted from a single-point to a two point system in order to achieve a more favorable orientation for a water landing. The landing system consists of three parachutes (drogue, pilot, and main parachute), two motors, reefing cutters, pyrotechnic cutters and separation assembly, disconnects, risers, and attaching hardware. The recovery system consists of a hoist loop, flashing recovery light, pyrotechnic cutter, dye marker, and flotation material.

Having discarded plans for a touchdown on land, the crews of Gemini were trained for spacecraft egress on water. Fortunately, as the flotation characteristics are a function of the center of gravity location, the Gemini capsule had only one mode of stability. Flotation of the Gemini capsule was achieved by displacing water with the cabin vessel and the equipment in the floodable bays. To improve the flotation attitude, additional flotation material (Styrofoam) was installed under the equipment in the side bays. Recovery teams would then attach a flotation collar around the capsule to provide added stability. Having a hatch directly overhead of each crewmember greatly facilitated emergency egress operations

Like Project Mercury, Gemini crews had to simulate water egress following splash-down through use of a water tank at Ellington AFB. Training was performed in various configurations, with or without suits, and with the capsule either floating or submerged. Afterwards, the boilerplate capsule was transported to the Gulf of Mexico, where egress training resumed with fully suited crew members and life rafts. As Gemini crews could also need to eject over water, flight crews were given instruction in parachute landings.

Though the experience and knowledge gained in the landing and post-landing phases of a space mission grew progressively since the days of Gemini, the climate of NASA also became much more risk averse following the Apollo Program. Thirty five years later, during the design of the PORT Phase Two egress test as part of the Constellation Program, it was deemed to be of unacceptable risk to conduct live egress training and testing in the open sea.

5.4 Egress Systems used for Apollo and Orion

The evolution of the Apollo side hatch was investigated in Section 3.6.1. Here, we describe the other systems on the Apollo spacecraft essential to egress: the seats, the up-righting system, and recovery systems, as well as variations that were incorporated on the Orion spacecraft. The Orion capsule was originally designed for a crew of six, reduced later in the design process to four, and the CMUS was modified to accommodate the greater mass of the Orion capsule. The docking hatch was also considered a mandatory secondary egress route, so accommodations had to be incorporated to make it so. One significant consideration was that the mission of Constellation was, partly at least, to support the International Space Station. As the station resides in a 51.6 degree inclined orbit, a launch from KSC places the ascent groundtrack over increasingly difficult sea conditions to higher latitudes along the Eastern Seaboard. The prior Apollo launches ascended due east and over waters much more benign. So the design team of Orion had to consider the much more challenging sea environments into which the spacecraft might abort into and design more robust systems or simply accept the risks that a successful launch abort may place the crew into a sea environment quite hazardous to the crew and rescue teams.

5.4.1   Seat and Impact Attenuation System Design

Upon landing on water, the CM will receive a peak deceleration from between 12 and 40 g's, depending on where in the wave the CM lands and the degree of horizontal velocity upon impact. Two methods were employed to reduce the shock loads experienced by the crew: an internal attenuation system and an external attenuation system.   The initial attenuation system consisted of eight struts that connected the crew couch to the spacecraft structure and were designed to absorb energy on landing.   The external attenuation system was incorporated after initial drop tests on water revealed that, though adequate impact attenuation was achieved by the struts, a significant amount of water leakage into the crew module occurred, causing it to sink within minutes of impact. To address this, the Apollo capsule was modified with a region of crushable structure, as shown in Figure 5.2.

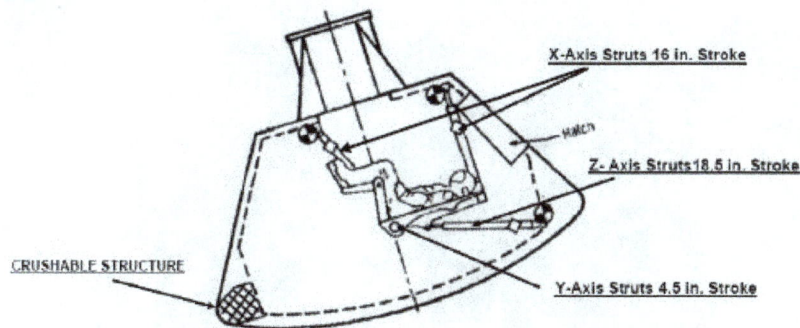

Figure 5.2 Apollo Couch Impact Attenuation System Configuration (McCullough, J., 2008)

As shown in Figure 5.3, the Apollo crew couch consisted of three seats connected together by a common pallet structure, which in turn was connected to the spacecraft by eight shock-absorbing struts. These couches support the crew during acceleration and maneuvers up to 30 g's forward, 30 g's aft, .18'g's up and down, and 15 g's laterally. A standard liner was employed across all three seats, rather than the personally molded type used onboard the Mercury spacecraft, and the headrest, back pan, armrests, and seat pan are all padded with removable 'Triloc' pads. However, even with these more favorable sustained loads, and the implementation of the crew couch and crushable structure, the impact experienced during landing was still much higher than expected. These high impact loads, coupled with the known fact that it was almost certain that the crew would be severely injured during a contingency terrestrial landing.

The Apollo seats were designed so that the seat pans could be configured to a position of 9, 85, 170, or 270 degrees. The 85-degree setting was used for launch and re-entry phases while on-orbit operations involving sleeping and tunnel activities required the seat pans to be configured to 170-degrees. The center couch can be configured to this 'flat out' position for egressing as all egressing through the docking hatch to the lunar module was performed from the center couch, and the lower armrests were removed and stowed, making easy egress from right and left couches into the center couch. The outer couches could not be configured to the 'flat out' position due to equipment interference beneath those couches. The left couch has two controller supports that also function as armrests. The right couch has only the inboard armrest. As shown in Figure 5.4, the armrests are held into position by a spring-loaded wedge into a slotted cam. The wedge is attached to a sleeve around the armrest, which can be lifted to position the armrest.

Figure 5.5 shows an individual seat of which the couch system is comprised of for both the Apollo CM and the Orion CM. Notice the lateral supports and positioning of arm and leg rests. The four seats that would be used in the Orion CM were equipped with head and torso supports that must be moved to facilitate egress. They would provide more lateral support, but its crew of four would find these additions adding further difficulty in an egress operation.

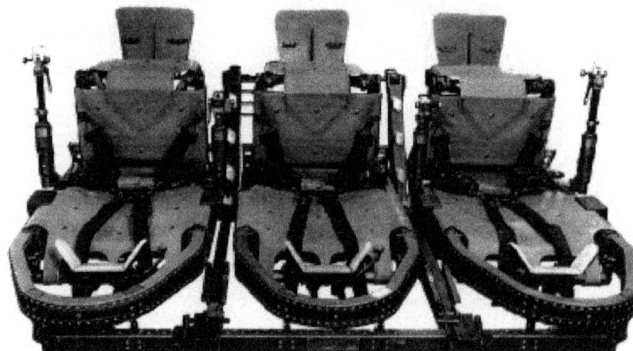

Figure 5.3 Detail of Apollo Crew Couch (Woods, D., and Brandt, T., 2009)

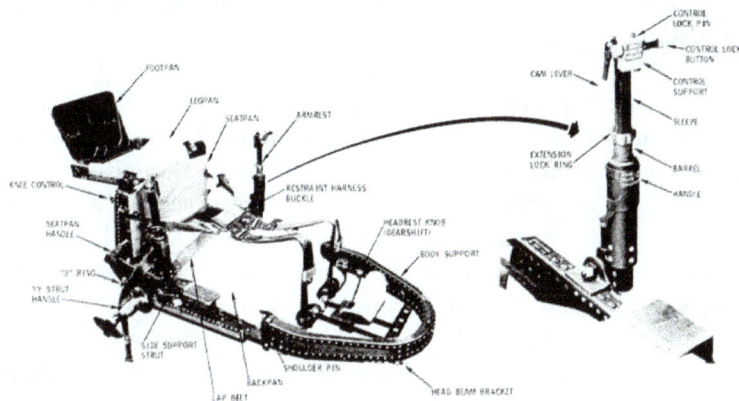

Figure 5.4 Apollo Foldable Couch Components (source: NASA Documents)

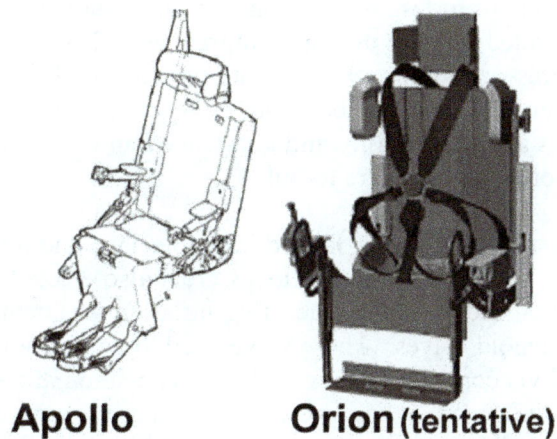

Apollo   Orion (tentative)

Figure 5.5 The Apollo Seat Unit (left) and the conceptual design of the Orion Seat Unit (right) with five-point harnesses and lateral supports. (source: NASA Documents)

5.4.2   Up-righting and Recovery Systems

The Apollo CM was optimally designed to support a three-person crew in spaceflight; it was not designed to be seafaring and, consequently, it had a nasty habit of preferring an inverted, "Stable Two" orientation as much as a right-side-up "Stable One" orientation. The Apollo design team needed to figure out a way both to prevent the CM from entering into a Stable Two and also a way to right the CM in the event it was to be in a Stable Two. Active means to control the CM splashdown was quickly deemed too complex and introduced too much weight as roll-reaction jets and torque motors would have to be added to the vehicle. A concept using a sea anchor was deemed unreliable since a constant tension would have to be applied to the anchor, which could not be guaranteed in the presence of ocean currents and winds. Further, though the sea anchor system might have been effective in preventing the CM from entering into a Stable Two, the system was eventually considered too complex and would have to go through an extensive qualification program.

Two methods of uprighting were considered: the first method lowered the center of gravity by manually adjusting the stroke of the couch system the full 16 inches so that the center of gravity could be lowered more than an inch, though this would not fully eliminate the possibility of Stable Two. The other approach was to create a moment about the center of gravity great enough to upright the CM. The energy techniques considered to achieve the torque required to right the CM from a Stable Two were rockets, sea anchor, water bag, aft compartment flooding, gas bag, and expandable foam bag. Of these options, only the gas bag options were deemed within the state of the art and within weight margins, and air compressor designs were considered the most technically mature and mechanically robust.

It was found that a single bag system would only cause the CM to list on its side, so additional configurations were tested. The final system, shown in Figure 5.6, consisted of three bags constructed of polyurethane-impregnated Dacron cloth that would be inflated by two compressors within a five minute time span. The bags were compressed into canisters that would release the bags once the compressors were activated. The additional components added reliability and guaranteed an effective inversion of the CM even if a compressor or an air bag were to fail.

Of the eleven Apollo splashdowns, six landed in Stable Two and the crew relied upon the uprighting bag system to invert. The Apollo 13 crew also elected to deploy the system, even though they had landed in Stable One. Bag inflation was manually activated by the crew through three solenoid valves. These valves had three positions (vent, fill, and seal) that could be selected via control switches or through an automatic sequencer if the flight was uncrewed. In the event a bag were to fail, crews were trained to reposition themselves to the aft bulkhead away from the failed bag in order to further lower the center of gravity (by up to an inch) and invert the CM.

35 years later, the Orion design also included a compressor driven righting bag system, though it was decided that a five-bag system would be the most reliable way of safeguarding against the CM remaining in a Stable Two. The Orion CMUS was designed off of the heritage Apollo system, through the three-bag Apollo system did not size up well to the greater size and mass of the Orion spacecraft, so an enhanced 5-bag system was designed, as shown in Figure 5.5. The Constellation Program required the CMUS system to right the capsule in less than three minutes to prevent cognitive deficits from setting in on a deconditioned crew and allow them to take action if required. Furthermore, crewmembers may become injured if they must release themselves from the restraints and the injury risk is significantly increased when the vehicle is up-righted while the crew is not restrained. If the crew is exposed to the stable-two posture for more than 15 minutes, the rescue will have to be treated as a medical emergency.

Figure 5.6 Apollo Uprighting Bag System (source: NASA Documents)

Physiological effects of Crewmembers Suspended in a Stable Two Orientation

Symptoms of a crewmember suffering from cognitive impairment due to being suspended start with general malaise, progressing to intense sweating, nausea, dizziness, hot flashes, brain function impairment that quickly worsens, respiratory difficulties, tachycardia, and progressively worsening arrhythmias, followed by a sudden increase in blood pressure and loss of consciousness. Deconditioning can complicate the situation as it produces susceptibility to orthostatic intolerance, muscle weakness, and uncoordinated movements. Additionally, there is a reduced physiological reserve and the crew is more vulnerable to injury.

5.4.3   Docking Hatch and Associated Egress Systems

As the Apollo CM was designed to perform rendezvous and crew transfer operations in space, it was logical to place the transfer hatch in the front of the vehicle, where it would be symmetrical and intuitive from an operations standpoint. It is not in an ideal location for ingressing the vehicle prior to launch, pad egress operations (the LAS system partially obstructs it), and certainly for post-landing egress operations. A side hatch is the most sensible ingress and egress route. But a redundant egress path, like any other mission critical system, is simply smart design. So, although the docking hatch serves a primary purpose during in-space operations, it also provides a secondary egress path in post-landing operations. The value of this secondary egress path was recognized in the Apollo program and defined in the requirements of the Constellation Program, and certain performance requirements had to be met concerning emergency egress out of the docking hatch.

Apollo Docking Hatch

The Apollo CM docking hatch, shown in Figure 5.7, is located at the forward end of the CM ingress and egress tunnel and served to both retain pressure and protect from external heating as provide an egress tunnel to perform EVA and for crew transfer to the LEM. The hatch is removable only into the crew compartment and is retained at the forward end of the CM tunnel by six separate latches joined by a linkage to a central drive ring that is driven by an actuating handle within the crew compartment. The latch position is controlled either by a gearbox assembly from within the CM or by a tool interface drive shaft from outside the CM. This handle is designed so that the hatch can be opened by a single crewmember using only one hand to sweep the handle up to 80-degrees. In addition, the latches may be moved from the unlatched to the latched position by disconnecting the gearbox and by using a special tool to drive a pinion gear that meshes with the drive ring. The gearbox is equipped with a single activating handle that is designed so that a single push stroke is required to latch or unlatch the hatch. The hatch

also contains a pressure equalization valve. Prior to hatch removal, the pressure needs to be equalized through this valve, which can be activated either from within or from outside the CM. Because the diameter of the hatch is greater than the CM tunnel seal retainer, the hatch can only be removed toward the interior of the CM.

Figure 5.7 Apollo Docking Hatch (credit: Kansas Cosmosphere)

Orion Docking Hatch and Docking Hatch Winch (DHW)

The Orion Docking Hatch was designed to open inward (into the cabin) and is pressure sealed. During on-orbit operations, the hatch is removable and storable as to not interfere with operations while docked with the ISS or other spacecraft. Located at the end of a 30" tunnel of 38" diameter that rises above the habitable crew space, the 34" diameter circular hatch weighs 138 lbs and requires the use of a docking hatch winch to raise or lower the hatch. The raising or lowering of the hatch in a one-G environment is considered a contingency operation and can be performed once, in preparation for an emergency egress.

The docking hatch of Orion differed significantly from that of Apollo because the Constellation Program required a timely post-landing egress operation be possible through it. Orion had no requirement to allow for resealing of the hatch ad, as hatch activation is considered a part of an emergency egress operation; the only requirement is to provide an egress path in a 'reasonable' amount of time. The ability to access the crew through the docking hatch by a rescue team with minimal training was considered highly desirable.

The Orion Docking Hatch Winch was a lightweight system that was proposed to be rapidly configurable to both lower the docking hatch, if the vehicle is stabilized in a Stable One orientation, or also to hoist upwards the docking hatch in the event the CM was stuck in a Stable Two orientation caused by a failure of the CMUS system. The Docking Hatch Winch (Figure 5.8) could then be used to pull the hatch upwards, assisted by its natural buoyancy. The winch itself was a simple device that could deploy or retract webbing depending on the direction the crank was turned. Figure 5.8 shows how the nylon webbing could be configured through both the hatch and the winch in order to lower the hatch safely and also how the same winch and webbing would be used to lift the docking hatch upwards in the event the CM was stuck in a Stable Two orientation and the crew was forced to egress. Keep in mind that the system had to keep the hatch controlled while under descent, lest the 138lb hatch act as a bell striker, agitated by sea motion, moving about the CM with the crew trapped inside!

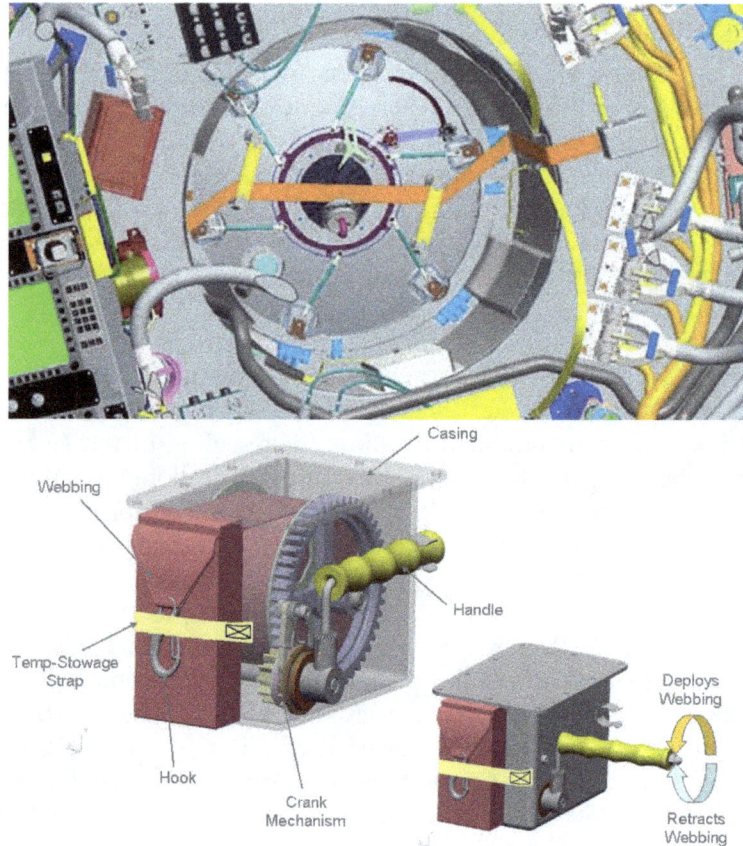

Figure 5.8 Orion Docking Hatch Winch (credit: NASA Documents)

5.4.4   Orion Egress Blanket

Most post-landing emergency egress operations are expected to occur upon or shortly after landing. The Orion capsule considered the docking hatch a viable secondary egress path, provided it was possible for the crew, especially a deconditioned member, to ascend to the pitching and rolling hatch and egress down the side of the spacecraft into the water and eventually into the emergency raft. But the capsule has just landed from a fiery decent and the exterior of the spacecraft will still be extremely hot from the aerothermal heating. It is likely that, in an emergency, the crewmembers could slide down the outside of the CM, but the pitching of the capsule and the unpredictable currents could create an extremely hazardous situation. The crewmembers should descend, if possible, in as controlled a manner as possible. The Orion CM was designed with an egress blanket so that crewmembers could rappel down the side of the CM while not getting burned by the residual heat. As shown Figure 5.9 for the side hatch and the docking hatch, the egress blankets are lightweight, thermally insulated, and easily deployed. A descent line is provided so that crew members can descend in a controlled manner and best assess the currents so that they might best ingress the raft directly and not be stricken by the base of the CM as they enter the water.

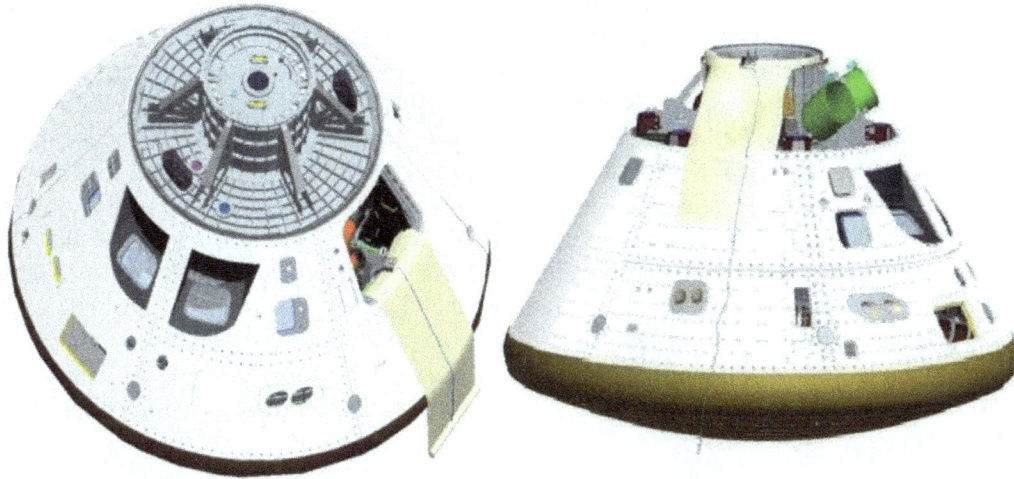

Figure 5.9 Egress blankets for side hatch (left) and docking hatch egress (right) (credit: NASA documents)

5.4.5   Rescue Aids

Rescue teams have been able to provide equipment to facilitate and reduce the risk of a rescue operation. For a nominal rescue of a water-landing capsule, a first priority is to stabilize the CM as much as possible. The attachment of a flotation collar, as shown in Figure 5.10 for the Apollo 11 CM, serves to stabilize the CM in terms of its oscillatory behavior and its tendency to invert itself. The addition of a sea anchor further stabilizes the CM with respect to the current of the water. Both the flotation collar and the sea anchor are typically attached by a rescue team prior to opening a hatch. Obviously, in the event of an unassisted emergency egress scenario, the crew would have to manage without the flotation collar and sea anchor.

Figure 5.10 The floatation collar attached to the Apollo 11 CM (credit: Smithsonian Institution)

115

External egress aids are often provided to facilitate the operations that are to be conducted by the rescue teams, though these must be designed so as not to drive vehicle mass or be vulnerable to damage through the high degree of aerothermal heating that they would experience during reentry. For daytime visual acquisition, sea dye was provided to the crew that could be remotely activated through a 'DYE MARKER' switch inside the CM, which is collocated with the uprighting bag switches, recovery beacon switches, and the swimmer umbilical deployment switch, which deploys an umbilical to assist the rescue team initiate the rescue procedure. The rescue team can attach a headset through the swimmer umbilical interface to communicate with the crew from outside the CM. Finally, any equipment that may provide a risk mitigation benefit by providing redundancy is always desirable, and anything that can be provided by ground teams, either prior to launch or after landing, incurs no vehicle mass penalty. A back-up capability to provide a critical capability (e.g. uprighting from Stable Two) may be a capability that a rescue team could be trained to provide, increasing vehicle reliability without driving vehicle mass.

5.4.6 Soyuz Hatch

The Soyuz DM has only one hatch – the one that connects to the OM while in orbit and is sealed before the OM is discarded prior to re-entry. From inside, the hatch can be manually opened by rotating the lever that serves to retract six retaining rods, as shown in Figure 5.11.

Figure 5.11 Soyuz docking hatch (source: Kansas Cosmodrome)

5.5 Space Shuttle Egress Systems

The Space Shuttle could support up to seven crewmembers – a lot more than the three supported by both Apollo and Soyuz. Several egress methods were provided to the crews of the Space Shuttle including egress through the side hatch or egress through a top hatch, called 'Window Eight'. After the Challenger accident, an additional egress method was provided so that crews could bail out the side hatch while in flight, termed a 'Mode VIII' egress. Here we detail the Shuttle seats, the side hatch and the 'Window Eight' egress path, egress aids such as the emergency egress net installed for facilitating a launch pad emergency egress and the Sky Genie, a system used to facilitate Window Eight egress operations.

5.5.1   Space Shuttle Seats

The Space Shuttle provided up to seven seats for crew positioning and restraint during all phases of flight, as shown in **Error! Reference source not found.**. Two seats are allocated to the pilots and five that can be used by Mission Specialists and Payload Specialists. Each seat was designed to accommodate a fully suited crewmember, cooling unit, and stowage bag. The seats secured a crewmember with a five-point harness fastened with a rotary buckle and also secured oxygen hoses, communications, and power cables for launch and entry. The seats are electrically powered for horizontal and vertical adjustment and equipped with mounting provisions for Rotational Hand Controllers (RHCs). All seat backs have two positions: 2° forward (launch position) and 10° aft (entry position). The recumbent seats are in the 10° aft position for entry.

The Mission Specialist and Payload Specialist seats are equipped with quick disconnect fittings that interfaces with mounting studs on the floors of the flight deck and middeck so that the Mission Specialist seats could be removed, folded, and stowed for on-orbit activities. All Mission Specialist and Payload Specialist seats can be fitted with an emergency egress step for use when egressing the vehicle via the escape panel on Window Eight. The step can be extended or stowed for launch and entry.

5.5.2   Space Shuttle Side Hatch

The 300 lb, 40-inch diameter side hatch, shown in Figure 5.13 from the interior, is the primary means of entering and exiting the orbiter. The hatch has a 10-inch diameter optically pure window in the center and is attached to crew cabin structure by hinges and secured closed by 18 hatches. The hatch requires up to 13 lfb to unlock, 16 lbf to rotate, and 25 lbf to overcome vent position detent. When the vehicle is in a horizontal position, the hatch opens outwardly 90°. During normal operations, the flight crew does not need to operate the side hatch. However, crewmembers may need to open the side hatch for a prelaunch or post-landing emergency, as well as part of a Mode VIII egress.

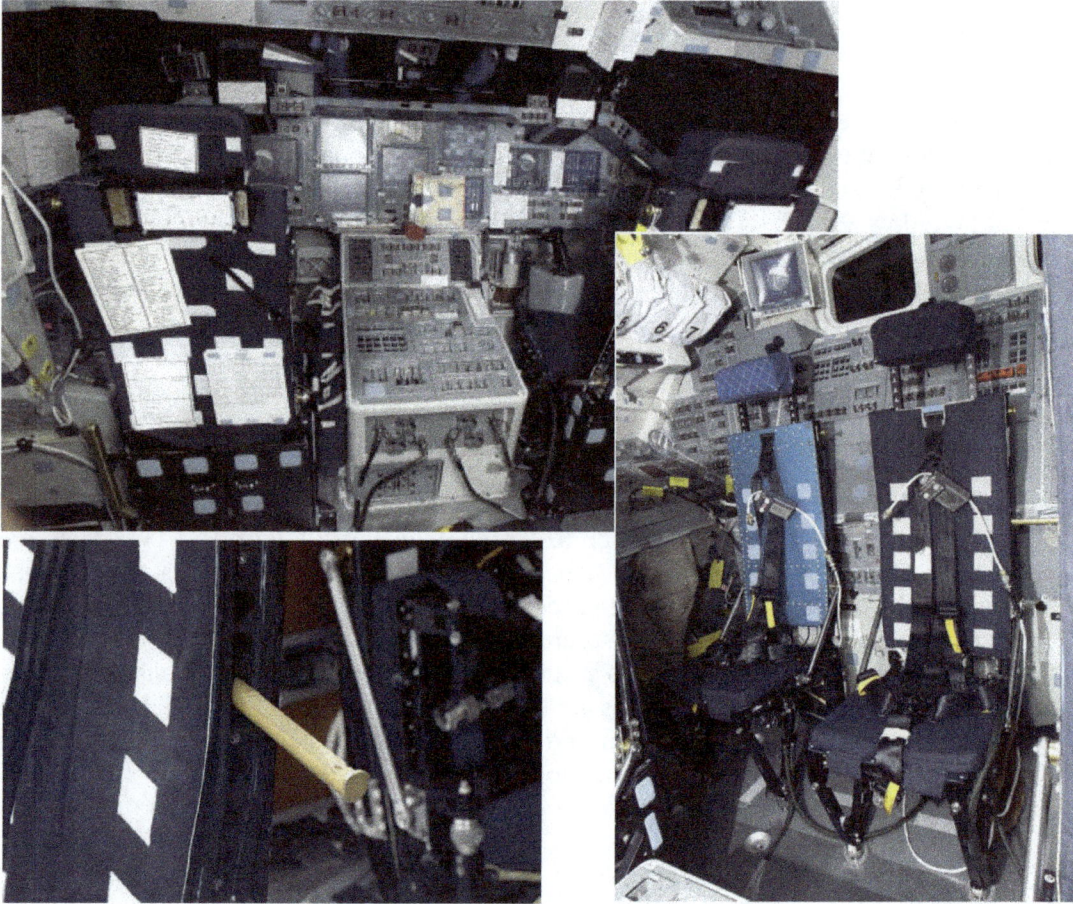

Figure 5.12 Space Shuttle seats. Commander and pilot above , Mission and Payload Specialists lower left, egress step lower right. (credit: United Space Alliance)

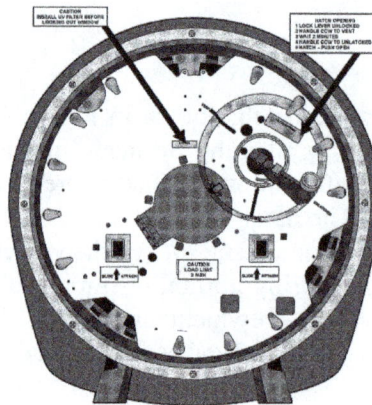

Figure 5.13 Space Shuttle Side Hatch (credit: United Space Alliance)

5.5.3 Space Shuttle Emergency Egress Net

The Emergency Egress Net, shown in Figure 5.14 and commonly referred to as the trampoline, serves as a platform for crew emergency egress when the orbiter is oriented vertically on the launch pad. With the removal of the internal airlock, a large volume was created behind the two seats on the middeck. To allow quick and safe crew egress during an emergency, the trampoline is installed prelaunch by ground crews. The trampoline is installed in the middeck, parallel to the aft bulkhead. Its span is from the deck to the ceiling and from the interdeck access ladder to the aft lockers. Four closeout nets are installed around the trampoline to keep items from falling to the aft bulkhead. The brown mesh closeouts and are located at the ceiling, deck, and port side of the trampoline and are attached to crew module structure using Velcro.

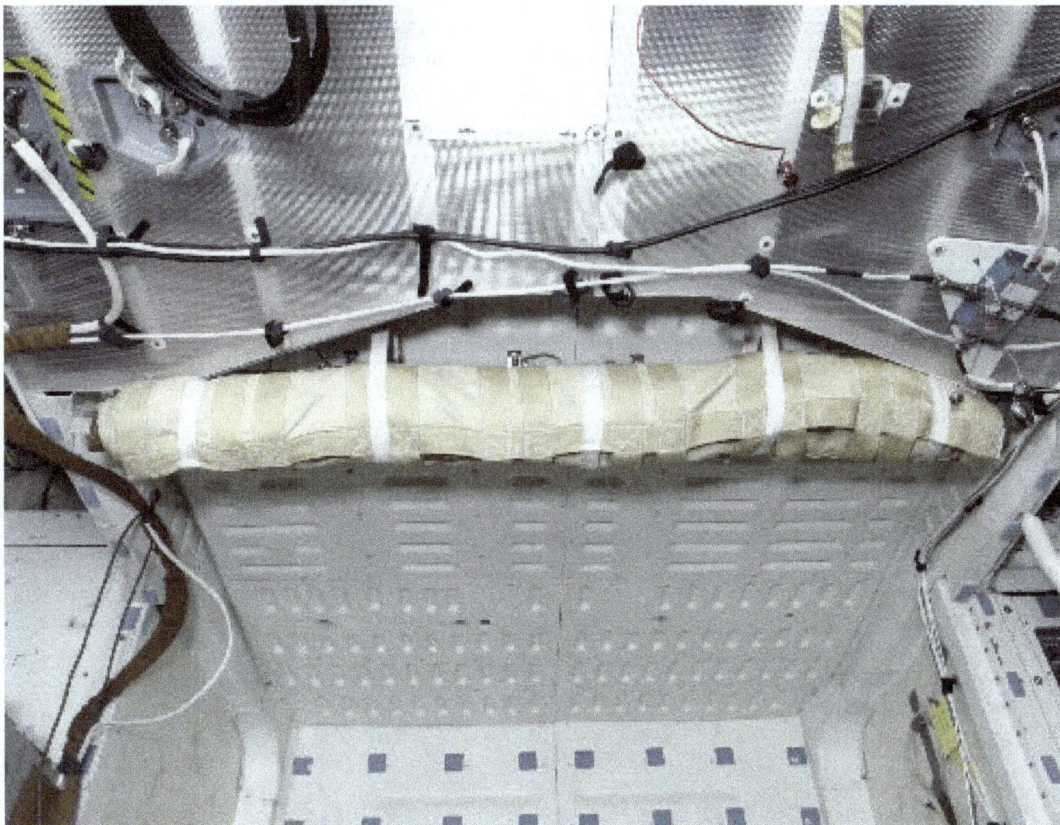

Figure 5.14 Space Shuttle Emergency Egress Net (credit: United Space Alliance)

119

5.5.4   Space Shuttle Window Eight Escape Panel

If emergency egress through the side hatch is impossible, the crew has an alternate route available, which is via the Window Eight escape panel. To use this secondary route, crewmembers:

    a. Jettison the Window Eight escape panel using the panel pyrotechnic system
    b. Attach to a Sky Genie (see next section).
    c. Egress through Window Eight by climbing atop the Mission Specialist seat

The relation between the window and the seats is depicted in Figure 5.15. The pyrotechnic system for jettisoning the escape requires no orbiter power to operate and is initiated by the jettison ring handle, located on top of flight deck center console. The escape may also be jettisoned by the ground crew via an external handle on starboard side of the orbiter forward of the wing. The crew then descends down the starboard side of the orbiter by using a 'Sky Genie'.

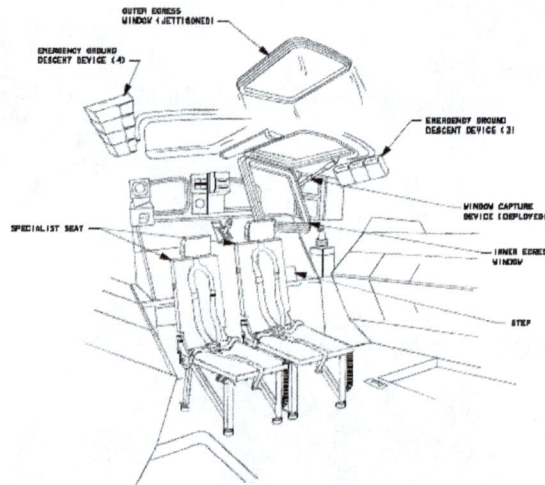

Figure 5.15 Window Eight Egress Components (credit: United Space Alliance)

The Sky Genie

If the Shuttle Orbiter side hatch fails to jettison, or if the egress slide fails, the flight crew must jettison the overhead emergency escape panel and use the sky genie descent devices to lower themselves to the ground. The SKY GENIE® is a simple device used to aid crew descent in the event of an egress out of a port, such as window eight, that does not allow access to the ground. The crew could then rappel down the side of the Orbiter. The Sky Genie operates on the principle of friction created by turning a nylon line around a metal shaft within the unit, as shown in Figure 5.16. The speed of descent depends on the number of turns around the shaft and the weight of the crew member, so the device must be calibrated for the average weight of a suited crewmember.

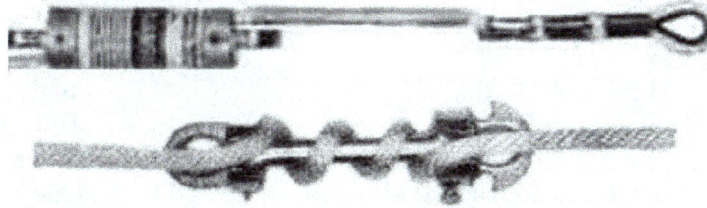

Figure 5.16 The SKY GENIE® (source: Descent Control, Inc.)

5.6 Emergency Post-Landing Survival Kits

Due to the risk of landing somewhere besides the primary landing zone, either by malfunction of the re-entry system or through an on-orbit contingency that requires an immediate de-orbit, space crews have always been provided an emergency survival kit with provisions to support a crew trained in survival techniques sufficient time to be rescued by contingency rescue teams.

Survival kit composition always involves a trade study of the most likely environments the crew might be faced with, the duration the crew might be expected to wait before rescue, and the associated risks of the environment. As each element of a survival kit contributes to the overall vehicle weight, it is essential that everything included in the kit has a reasonable justification. Gear required for arctic survival would not be needed for an orbital flight of low inclination, though might be essential for a mission to a highly inclined orbit where the groundtrack expands to high latitudes.

Detailed here are the survival kits as were provided to the crews of the Mercury spacecraft, the Soyuz spacecraft, and the Space Shuttle. The Russians clearly place a high emphasis on post-landing preparation in the event of contingencies, but this is likely due to the generally higher inclined orbits the Soyuz spacecraft are launched into, as well as the history of Soyuz 5, where the lone cosmonaut landed in the Ural Mountains after an erroneous re-entry. His ability to survive in the frigid conditions long enough to seek shelter and later rescue was a tribute to the planning and preparation of his emergency survival kit. It is also likely that, because he was tracked by wolves while seeking shelter, pistols and ammunition were included in Russian survival kits ever since.

5.6.1   Project Mercury Survival Kit

There was no room for the astronaut to move about in the Mercury capsule; the capsule itself was literally built around the pilot. To the astronaut's left was placed his survival kit. To his right were placed storage facilities for water, food and waste. A flashlight was placed adjacent and to the left of the observation window. The survival kit placed in each Project Mercury spacecraft contained the following items:

1.	PK2 life raft	10.	Survival ration
2.	Water container	11.	Matches
3.	8 pint desalinating kit	12.	Whistle
4.	Shark repellant	13.	10 feet of nylon cord
5.	Three sea dye markers	14.	One tube of zinc oxide
6.	First aid and medical kit	15.	One bar soap
7.	SARAH rescue beacon	16.	Sun glasses
8.	Signal Mirror	17.	One pocket knife
9.	AN/PRC-32 radio		

A knife and flashlight are attached to the pressure suit, which was designed to float unless it became filled with water, as had happened with Gus Grissom during his MR-4 flight

5.6.2 Apollo Portable Survival Kit

Apollo survival equipment is contained within two cloth rucksacks, as shown in Figure 5.17. Equipment contained within the two rucksacks are designed to support the crew for 48-hours after landing. It was expected that the rescue teams could arrive within this time for any credible contingency.

The first rucksack contains a water-resistant, hand-held UHF beacon and transceiver. In beacon mode, the transceiver can operate for a period of up to 24-hours. Also inside this rucksack is a Survival Light Assembly that contains three compartments, the first housing a flashlight, the second an emergency strobe for night signaling, and the third a compartment containing a fish hook with line, a fire starting kit, a mirror, needle and thread, a compass, and a whistle. Also included is a desalination kit, which includes a desalination bag with tablets that will produce one pint of fresh water per tablet. Two survival knives are included as with three sets of polarized, plastic sunglasses and two containers of sun lotion. Three aluminum cans are provided as well, each containing five pounds of water (5/8 gallon).

The second rucksack contains flotation gear including a three-man life raft with an automatic CO_2 inflation system, a sea anchor, dye marker, lanyards, and a sunbonnet for each crewmember.

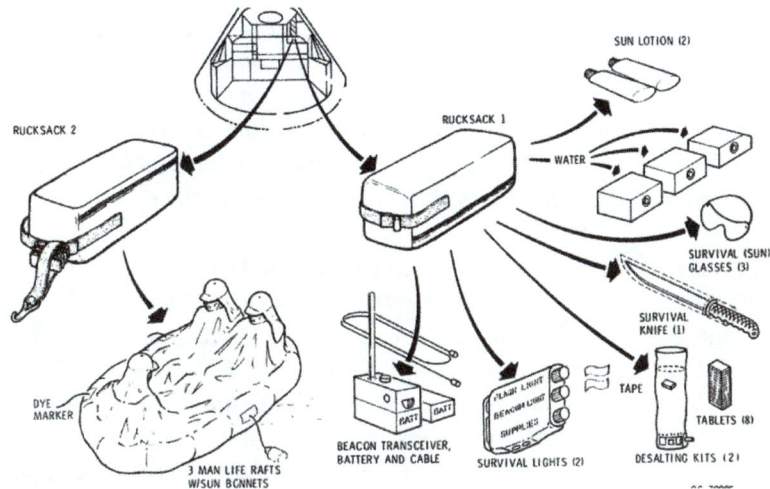

Figure 5.17 Apollo Survival Kit Components

5.6.3 Soyuz Portable Survival Kit

The 32.5 kg Portable Survival Kit provided in the Soyuz DM is extensive and comprehensive. After the ballistic return of the TMA capsule containing the ISS-6 expedition crew of Bowersox, Pettit and Budarin, a satellite phone system has since been added to the kit.

The book *Russia's Cosmonauts* provides some details of the Soyuz Portable Emergency Kit:

> *Every Soyuz craft carries a Granat-6, «Гранат-6» (Pomegranate) survival pack, which includes a "Forel" («Форе», Trout) hydro-suit – a one-piece orange nylon flotation suit with attached rubber soled feet and a hood trimmed with "CCCP". The suit contains a "Neva", «Нева» inflatable collar with an emergency mouthpiece, emergency beacon and a signal device on the shoulder. It also has rubberized cuffs, Velcro-close pockets on the legs (with ten pairs of small rings on the legs and eight pairs of grommets on the boots), and a pair of brown jersey mittens with separate thumb and index finger stalls, with watertight cuffs and adjustable orange nylon wrist straps. There is also a TZK-14 cold weather suit, with a royal blue nylon zip front anorak with attached mittens. This has two slash pockets with contrasting zips and a draw closed waist. Also included is a wool knit balaclava, a lined wool knit cap with button flaps, wool gloves, one pair of socks and one pair of nylon over boots, elasticized at the top with Velcro-close at the heels. There are three other orange nylon packages in the pack. These contain survival equipment including a large canteen, a soft flask, dried food, a medical kit, a frying pan, signals and flares, a machete (which also doubles as the shoulder rest of the rifle/shotgun), a Makarov pistol*

with cartridges (TP-82m), a foraging bag, fishing tackle, and metal wire garrotes for use as a saw as well as for hunting. The combination of the "Forel" suit and thermal suit is intended to keep the wearer alive for up to twelve hours, if needed, in water of 2°C, with an ambient air temperature of -10°C (14°F). Coupled with the shelter of the descent craft, it is hoped that the clothing and supplies could support a cosmonaut for up to three days in conditions of severe cold.

The Soyuz Portable Emergency Kit is contained in two orange triangular aluminum containers with grey canvas tops and zippers, as shown in Figure 5.18. They are stowed between the seats. The NAZ-3 is designed to serve the three cosmonauts for 72 hours and includes (from left to right, top to bottom in photo): one Makarov pistol and ammunition; onewrist compass, 18 waterproof matches with striker; machete; fishing kit; strobe light with spare battery; 8 fire starters; folding knife; an antenna; 3-pair wool gloves; one signal mirror; one NAZ-7M type medical kit; a penlight; one R-855-YM or R-855-A1 radio; two "Priboy 2S", ПРИБОЙ-2С radio batteries; and three wool balaclava hoods. In addition to what is shown in the Portable Emergency Kit, there are also available three ПСНД hand-held flares; one 15 mm flare set; a whistle; a sewing kit; insect repellent; a lightweight wire saw; rations and one 2-liter water container.

Figure 5.18 Soyuz Portable Survival Kit with Immersion Suit

5.6.4 Space Shuttle Crew Survival Gear

Space Shuttle crewmembers are provided an array of emergency rescue/survival gear, stowed in both leg pockets on their flight suits (Figure 5.19). The rescue/survival gear in left leg suit pocket includes: 1) One 'Pengun' Flare Kit (MK-70) and seven cartridges, 2) two green 12-hr, 'Cyalume' lightsticks, 3) one MK-124 day/night flare and smoke signal,

3) one strobe light, 4) one exposure mitten (the other mitten is in right leg pocket). The rescue and survival gear in the right leg suit pocket includes: 1) one PRC-112 survival radio (24-hour battery life), 2) one signal mirror, 3) two motion sickness pills, 4) an exposure mitten, and 5) a shroud-line cutter/knife. Also included is a parachute harness, a LPU for egress into water, and two emergency breathing apparatus (EBA) bottles that will provide oxygen for up to 10 minutes at sea level. Two liters of emergency drinking water is also supplied near the EBA bottles.

Rescue and survival gear is stowed in the life raft package (Figure 5.20) which includes a one-person survival raft with spray shields that can be inflated by one of two CO_2 cartridges or through oral inflation. Tethered to the raft is a bailing cup, bailing pump, and additional sea dye marker. A SARSAT personal locator beacon (PLB) with a 24-hour battery is also provided.

Figure 5.19 Space Shuttle Rescue and Survival Gear (credit: United Space Alliance)

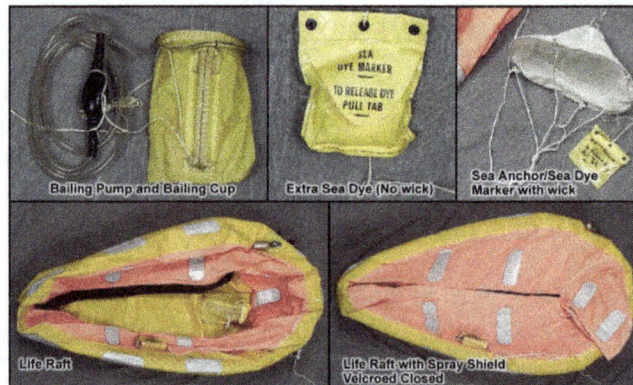

Figure 5.20 Space Shuttle Liferaft Package (credit: United Space Alliance)

Case Study: Soyuz 5

In January 1969, less than two years after Cosmonaut Korolev was killed in Soyuz 1 following a malfunction of the parachute system, Boris Volynov faced a similar fate onboard the Soyuz-5 vehicle. After docking with the Soyuz-4 spacecraft to demonstrate the capability to rendezvous and dock in orbit, two of the three Soyuz-5 crew members were transferred to the Soyuz-4 vehicle and Volynov was returning to Earth alone.

Nominally, Soyuz spacecraft shed the forward-mounted OM and the aft-mounted service module before re-entry. The DM that secured the crew and parachute system would re-enter in a stable orientation with its heat shield 'down'. But when the pyrotechnic bolts had fired on Soyuz-5 that were to separate the DM from the OM the OM failed to fully separate and Volynov could see the whip antennas from that module still extending past the window. If unshielded portions of the DM were subjected to the 5000 degree (Celsius) heat of reentry, the vehicle and its pilot would disintegrate. But the DM was already in the hands of Newton, with the OM blocking the heat shield and the vehicle beginning to tumble, and there was nothing that either Volynov or Mission Control in Moscow could do.

Realizing that the rescue would be hours away, Volynov quickly surmised that he would not last that long in the Siberian winter. As fate would have it, he spotted smoke rising from the hut of a local family. It was noon local time, the sky was clear, and Volynov headed to the hut wearing anything he could to preserve heat. Hours later, the rescue team arrived by helicopter expecting to encounter a scene that haunted the minds of all who would recall *Soyuz 1,* but found instead an empty DM and a trail of blood where Volynev had spat that led to the hut where a recuperating Volynev was waiting.

Volynov made a time critical decision after landing. Spacecraft crews are trained to survive in a variety of environments and this training probably gave him the intuition to react appropriately. He estimated the time a rescue effort would arrive, and the time he would have had to avoid freezing as the sun descended towards the west. He made the correct decision to seek out the distant vertical line of smoke in the sky.

As the vehicle tumbled into the atmosphere, the fuel tanks exploded as the exit hatch bent inwards as the rubber seal surrounding it began to smoke. The cabin was growing toxic and, as Volynev was not wearing a space suit, the temperature was becoming exceedingly hot inside the capsule. Realizing he had mist likely seconds to live, he quickly tore the most recent pages from his log and tried to secure them as deep in his jacket as he could; perhaps somehow his body might be found to protect those last few entries. But then, all of a sudden, the OM broke loose and the DM abruptly settled itself into a heatshield 'down' orientation.

The reentry was well off target, having carried the extra mass of the OM through a large portion of the reentry. It was uncertain whether the parachute system would still work. But the parachutes managed to deploy and though the parachute lines managed to partially tangle, Volynov landed 2000km off target on the snowy Ural Mountains near Orenburg, Russia. The force of the landing, exaggerated by a partially closed parachute canopy and the failure of the landing rockets to fire, dislodged Volynev from his seat and knocked out several of his top front teeth. Exhausted and bleeding, he realized that he was still not out of danger. Replacing the immense heat of reentry, now a bitter cold seeped into the DM; it was minus 38 degrees Celsius outside (-40 degrees Fahrenheit).

In addition to Soyuz 5:

Excessive heat can become a quick danger if the crew is unable to egress or otherwise cool the capsule. On 17 June, 1988, Soyuz TM-4 landed in a heat wave that recorded temperatures in excess of 42°C. Snow can obstruct communications, as the crew of Soyuz-24 found on 25 February 1977 when the DM was lost to rescue helicopters because the SAR antenna had jammed closed due to the weight of the rapidly accumulating snow. Communication was established only after the crew was able to egress and clear the antenna manually. Later, the Soyuz T-7 crew found themselves landing in a similar snowstorm, much more intense that the weather forecasts predicted, and recovery forces could not safely land in the limited visibility conditions near the landing site. And then the Soyuz-15 capsule descended through an active thunderstorm on 28 August 1974; there was risk that convective activity could obstruct and collapse the landing parachute system.

6   Egress Procedures and Operations

Egress procedures and post-landing contingency planning has evolved greatly as the collective experience of crewed re-entry, landing, and rescue and recovery operations has matured. Yuri Gagarin's Vostok 1 required the astronaut to egress before landing; the Soviet space program electing to have the cosmonaut conduct a proven operation rather than attempt a crewed landing. In contrast, more recent space programs require the crew to stay in the spacecraft until rescue unless presented with a dire post-landing contingency.

As spaceflight experience increased, so did knowledge of the possible contingencies that could happen during the landing and post-landing phases. Some of this experience was gained through the misfortunes of the crew of Mercury MR-4 and Soyuz 23, both of which were suddenly faced with life-threatening post-landing scenarios, or through failures of critical but redundant systems, such as the Apollo 15 parachute malfunction. Other knowledge was gained through the use more thorough analytical methods as well heritage design of individual components essential to the landing and post-landing phases of flight.

6.1   Mercury Spacecraft Egress Procedures

The Mercury emergency egress committee was headed by Astronaut Gordon Cooper to insure maximum safety for the astronaut and minimum danger to the launch support personnel and the rescue teams. From the initial development of the Mercury capsule, the side hatch provided the only means to ingress or egress the capsule. This hatch was held in place with locking cams that were connected to the release handle by mechanical linkage and opened easily through the following procedure:

1. Remove harness and personal leads
2. Detach survival kit from capsule
3. Depress handle lock release button
4. Pull release handle down
5. Push hatch out of capsule
6. If the capsule was in the water, the survival kit and raft were pushed out of the hatch first so that the astronaut could use the inflated raft for support.

The seven Mercury astronauts then used the hydrodynamics tank at the Langley Research Center to validate the hatch design and to practice egressing via the side hatch in smooth and artificially induced rough water. A second phase of testing was conducted in the Gulf of Mexico where the capsule was dropped from a ship into the water and the astronauts practiced egressing in rough waters. Four egress methods were tested:

1) The astronauts jumped into the water and then inflated their life rafts.
2) The astronauts inflated their life raft, and then jumped into the water.
3) A recovery helicopter would drop a hook to the astronaut waiting inside the floating capsule and lift the astronaut directly from the capsule.
4) The astronauts performed an underwater egress.

In these initial egress tests, no specific procedures were initially established for Project Mercury; each astronaut could choose which method of egress they intended to use from the four methods tested. The underwater egress was the most dynamic and risky option that was tested. This egress method involved the astronaut blowing the side door hatch and waiting for the capsule to fill with water. Once the capsule began to sink, the astronaut could swim out and up to the surface. In conjunction with the water egress training, the astronauts also spent about a day in a one man raft learning how to signal, protect themselves from the sun, and survive until the rescue forces arrived.

It was later determined that a second, more symmetric and thus more stable, egress path should be made available to the astronaut. Thus, in the interest of maintaining the most predictable stability characteristics in the water, the primary egress path to be used in the case an unassisted emergency egress needs to be performed was through the hatch on top of the spacecraft, where the parachute bay is located. This path was deemed less likely to induce a sinking situation, though water could still be quickly introduced into the capsule, especially in rough seas. To perform this egress, the astronaut would remove a small pressure bulkhead behind the antenna compartment, which was a time consuming procedure that would likely be impossible to perform if the astronaut was injured. For Alan Shepard's MR-3 flight, Shepard was advised to remain in the capsule and have access to recovery aids. But if it was necessary to egress the capsule, the preferred way was through the top of the spacecraft through the antenna compartment. These egress procedures, as presented in the Mercury MR-3 Flight Operations manual, are presented in Figure 6.1. Figure 6.2 and Figure 6.3 shows a representation of these procedures as were depicted in the operations manual and Figure 6.4 shows John Glenn performing an egress out of the top of the capsule. Removal of the side entrance hatch remained an alternate, though less desirable, egress route to be used only in the most time-critical emergencies, and astronauts were warned that, the capsule could quickly become swamped after the hatch is removed and that the hatch would sink below the waterline when burdened by the weight of the astronaut.

For Gus Grissom's MR-4 flight, the Mercury capsule was modified to include an explosively actuated hatch. Explosive removal of the side hatch provided a rapid egress route and was considered the preferable means to conduct an egress, though the egress route through the antenna compartment was maintained as a secondary egress route in the event that sea conditions made a side hatch egress undesirable. The side hatch was released by detonating a gasket type firing plate to sever the notched titanium hatch bolts and it can be actuated for egress on the launch pad as well as after landing by removing the cam-locked cap covering the initiator, withdrawing the safety pin, and striking the initiator button with the heel of a clenched hand (Figure 6.5). It is recommended the astronaut only detonate the hatch with a secured helmet and a closed visor to safeguard against the noise level of the hatch detonation. To facilitate a timely egress, the Right Main Instrument Panel may be removed.

SEDR 109-7(CL)

COOLING SYSTEM FAILS

1. Remove escape hatch and open suit.
2. If capsule overheats, evacuate capsule.

CAPSULE LEAKS AFTER LANDING

1. If leak is large, leave capsule.
2. If leak is small, check conditions outside before deciding to leave capsule.

EGRESS

1. Remove harness and disconnect leads.
2. Remove survival kit cover, insert knife and flashlight in kit pockets.
3. Remove right side of instrument panel and stow under main panel.
4. Sit on right side of capsule and release escape hatch.
5. Stow hatch on leg couch, step into seat, and climb into hatch opening.
6. Check survival kit secured to the suit.

E-11

SEDR 109-7(CL)

7. Disconnect electrical leads, and remove pip pins from parachute container.
8. Push parachute container and survival kit out of capsule.
9. Maneuver through the housing until shoulders are out of capsule.
10. Retrieve survival kit, inflate raft, and tie raft to capsule.
11. Get into raft.
12. Secure raft & start rescue procedures.

EMERGENCY EGRESS

1. Remove harness & personal leads.
2. Detach survival kit.
3. Handle lock release button - DEPRESS
4. Pull release handle down
5. Push hatch out of capsule

If Capsule Is In Water:

6. Remain in couch, push out survival kit, and inflate raft.
7. Use raft for support while leaving.

E-12

Figure 6.1 Project Mercury Egress Procedures

EGRESS

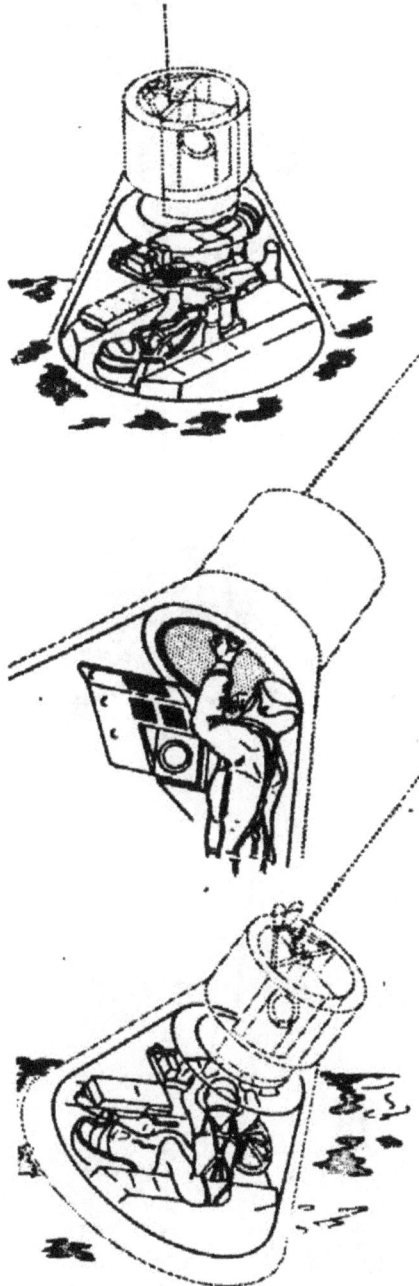

1. REMOVE HARNESS AND DISCONECT PERSONAL LEADS. THE HELMET MAY BE REMOVED IF DESIRED.

2. REMOVE SURVIVAL KIT COVER AND INSERT KNIFE AND FLASHLIGHT IN SURVIVAL KIT.

3. REMOVE RIGHT SIDE OF INSTRUMENT PANEL AND STOW UNDER MAIN PANEL.

4. SET ON RIGHT SIDE OF CAPSULE AND RELEASE ESCAPE HATCH. CAPSULE WILL START TIPPING WHEN ASTRONAUT SHIFTS WEIGHT.

5. STOW HATCH ON TOP OF LEG COUCH.

6. CHECK THAT SURVIVAL KIT LANYARD IS SECURED TO SUIT.

Figure 6.2 Mercury MR-3 Egress Procedures (1)

131

7 DISCONNECT TWO ELECTRICAL LEADS. REMOVE THREE PIP PINS FROM THE PARACHUTE CONTAINER.

8 PUSH PARACHUTE CONTAINER AND SURVIVAL KIT OUT OF CAPSULE.

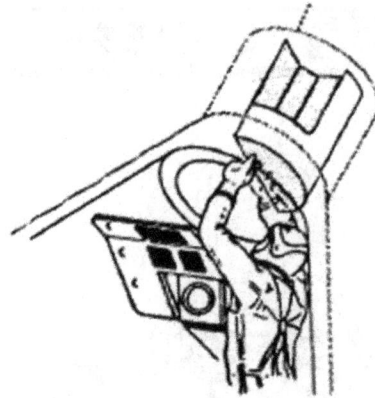

9 MANEUVER THRU HOUSING UNTIL THE SHOULDERS ARE OUT OF CAPSULE.

10 RETRIEVE SURVIVAL KIT, INFLATE RAFT AND TIE RAFT TO CAPSULE.

11 GET INTO RAFT.

12 SECURE RAFT AND START RESCUE PROCEDURES.

Figure 6.3 Mercury MR-3 Egress Procedures (2)

Figure 6.4 John Glenn during egress training (source: NASA)

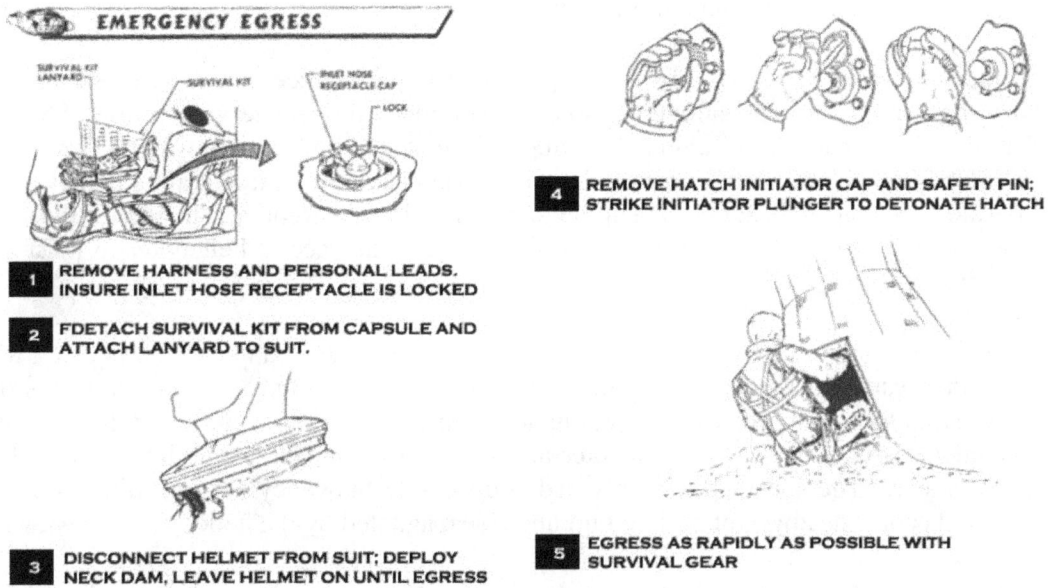

EMERGENCY EGRESS

1 REMOVE HARNESS AND PERSONAL LEADS. INSURE INLET HOSE RECEPTACLE IS LOCKED

2 FDETACH SURVIVAL KIT FROM CAPSULE AND ATTACH LANYARD TO SUIT.

3 DISCONNECT HELMET FROM SUIT; DEPLOY NECK DAM, LEAVE HELMET ON UNTIL EGRESS

4 REMOVE HATCH INITIATOR CAP AND SAFETY PIN; STRIKE INITIATOR PLUNGER TO DETONATE HATCH

5 EGRESS AS RAPIDLY AS POSSIBLE WITH SURVIVAL GEAR

Figure 6.5 Mercury MR-4 and later egress procedures (source: NASA)

Case Study: Liberty Bell 7

Astronaut Virgil "Gus" Grissom was America's second astronaut, and his suborbital flight onboard the "Liberty Bell" Mercury spacecraft proceeded flawlessly to a landing point only 3 mi (4.8 km) from the target point. The recovery operation was proceeding as planned when suddenly the side hatch jettisoned without warning.

> *"I was lying there, minding my own business,"* Grissom later commented, *"when I heard a dull thud."*

The side hatch was believed to have been activated during the recovery operation by an outside triggering of pyrotechnic valve when vehicle made contact with a squib-actuated cutter at the end of a pole that was being used to cut the spacecraft's antenna whip. After the hatch cover was ejected, The Liberty Bell 7 began taking on water and was sinking fast. Having unbuckled himself prior, Grissom could quickly egress the sinking capsule. He grabbed the instrument panel with his right hand and climbed out of the hatch.

Lieutenant James L. Lewis, pilot of the primary recovery helicopter, recognized that Grissom was safely away from the sinking capsule and continued the recovery operation of the capsule. But once Lieutenant John Reinhard, co-pilot of the nearest recovery helicopter, was successful in grappling the spacecraft's hoisting loop with the pickup pole, it had already picked up a lot of water. Seeing indications of engine strain, Lewis told Reinhard to retract the personnel hoist as he abandoned the capsule to sink and called the second helicopter to hoist Grissom from the water.

At first, Grissom's space suit kept him buoyant, but the air that had kept him buoyant while the helicopter team endeavored to save the sinking capsule was slowly escaping in spurts from his neck dam, and the situation was compounded by the fact that Grissom had failed to secure his suit inlet valve. The second helicopter arrived as Grissom was struggling to keep his head above water. The suit that had provided comfortable buoyancy several minutes earlier had quickly become deadweight pulling him under sea, agitated by the helicopter's downwash.

George Cox, of Alan Shepard's recovery team, quickly tossed the "horse-collar" lifeline to Grissom, who immediately climbed in and was hoisted to safety. In total, Grissom was only in the water for less than five minutes, but the last moments, encumbered by a saturated spacesuit, quickly endangered his life. Though his spacecraft was lost, Grissom lived to command the Gemini 3 mission.

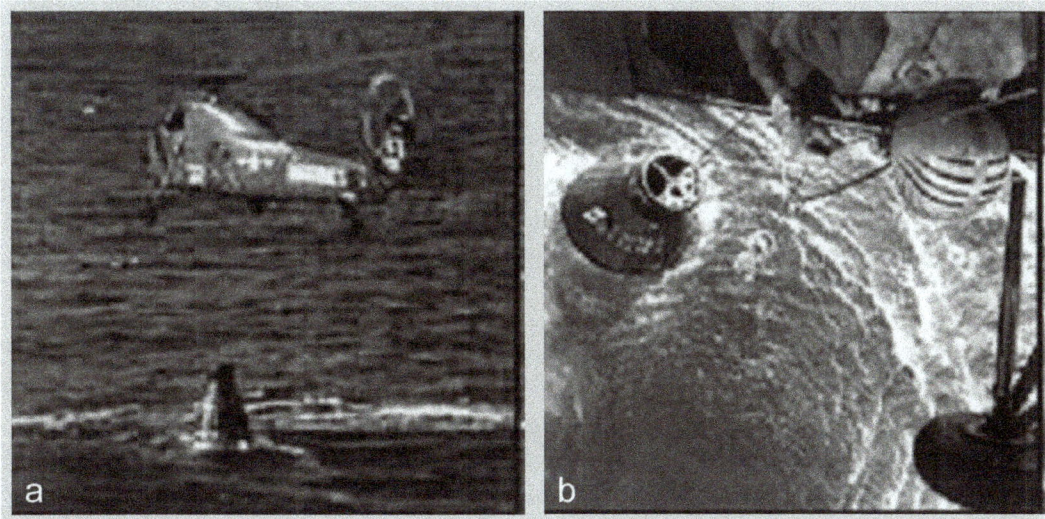

Figure 6.6 Liberty Bell 7 being hoisted, b) Liberty Bell 7 and its solo crew member, Virgil "Gus" Grissom, after being ejected into the water. (source: NASA)

6.2 Gemini Egress Procedures

In February 1965, the first Gemini crew, John Young and Gus Grissom, trained in a full-scale mock up in the turbulent waters of the Gulf of Mexico, fully suited up. Having learned more of the challenges and potential risks of the post-landing phase of flight from Project Mercury, a more methodical approach was taken to plan and train for egress in Project Gemini. Part of the Gemini astronaut's training involved escaping from the spacecraft while fully submerged, which was considered a real possibility.

In a nominal rescue and recovery operation, a recovery team would attach a stabilization collar to the capsule before extracting the crew, but in a contingency situation the crew might have to self-egress. Unlike the pyrotechnic hatch of Mercury, the two hatches of Gemini spacecraft had to be opened by hand and swung back on their hinges, allowing the crew to leave one at a time so as not to tip the rolling capsule in the event that the recovery crew was unable to attach a stabilizing collar. The crew then had to inflate and deploy their personal life rafts, each of which contained basic survival gear, lights, shark repellent and a radio transponder to aid location in an emergency. Crew members also had rubber neck dams that could be fitted over the collar of the space suit, effectively transforming it into a buoyant drysuit, preventing sea water from entering the spacesuit.

Egress procedures of the Gemini spacecraft were developed through a series of tests in 1964. The total egress development plan consisted of four test series:

1. Egress development using Boilerplate 201 in a calm water tank

2. Egress development using Boilerplate 201 in open sea
3. Static Article Number Five familiarization.
4. Evaluation and standardization of egress techniques utilizing Static Article No. 5.

Gemini crews practiced post-landing systems operations, suited and unsuited surface egress, use of the survival pack equipment, and helicopter pickup. Tests relied upon many measures to assure safety including use of a boom and crane, SCUBA gear, safety swimmers, and on-site medical doctors. The open sea provided a 'rough water' environment that produced 4 to 5 foot waves with occasional swells to 7 feet. Testing of the Gemini capsule was performed using a boilerplate capsule, labeled "Boilerplate 201". Boilerplate 201 mimicked the actual capsule in terms of weight, balance, and moments of inertia. The hatches, however, were similar to flight articles in form but were lighter and easier to open and close. Static Article No. 5 was a full engineering model that included all the details of a flight article. Details such as stowage of helmets and other loose equipment, donning of neck dams, acquisition of survival kits, actuation of life vests and rafts, and hatch operations were developed using such models.

Testing, as shown in Figure 6.7 and Figure 6.8, found that the spacecraft could roll to a total of 18° right and had buoyancy characteristics that favored a procedure where both crewmen would exit through the left hatch with the first crewman to egress staying on the right side of the spacecraft, draping himself over the right hatch, until the second crewman had egressed. This procedure was found to best prevent the possibility of the capsule flooding and sinking.

If the egress was more time-critical, both hatches could be opened and both crewmen could then enter the water forward of the hatches, but this was found to lead to sinking of the capsule in any sea state greater than calm and also drove the possibility of a crewman slipping and striking himself.

Egress procedures were also validated for an underwater egress as it was found that crewmen can escape from a flooded spacecraft sinking in any assumed underwater attitude, if the crewmembers held their legs together and pulled their bodies through the hatch opening with their arms. The buoyancy of the suit would then orient the astronaut towards the surface of the water.

Underwater egress tests were developed through five tests where the initial conditions had the vehicle in five different orientations. For each egress exercise with test subjects, the boilerplate was submerged with the left hatch open and the right hatch closed. Once the boilerplate was filled with water, opening of the closed hatch was not more difficult in any of the underwater test attitudes than opening of the hatch with the boilerplate in a normal attitude on the surface. The test subjects encountered no difficulty in getting clear of the boilerplate in any of the underwater test attitudes when the following procedures were followed:

EGRESS PROCEDURES AND OPERATIONS

1.   Keep legs close together and use legs only to push away from the seat.
2.   Keep torso and extremities away from protrusions.
3.   Pull the body through the hatch opening mainly by grasping the hatch opening edge and ingress/egress handle and pulling with the arms, keeping all other body motions to a minimum (except for pushing with the legs, if necessary).
4.   Use arm motions to swim clear of the spacecraft.
5.   Once clear, use the suit buoyancy to indicate the direction of the water's surface. It is possible to be disorientated and swim in a wrong direction.

Figure 6.7 Gemini Boilerplate 201 and Phase One Testing in Calm Water (source: NASA)

Figure 6.8 Gemini Egress Training in 'Rough Water' (source: NASA)

6.3 Apollo Egress Procedures

The Apollo spacecraft had more egress concerns than the Gemini vehicle. Whereas the Gemini capsule had hatches and ejection seats dedicated to each crewmember, The Apollo vehicle had to coordinate three crewmembers through one hatch. Furthermore, less predictable buoyancy characteristics complicated post-landing procedures. Egress training therefore required a high level of coordination between the crewmembers.

Apollo crew egress training was first conducted in a post-landing environment in water and then on the launch pad. The water egress training was conducted in three phases. The first phase included training on each of the components of survival gear that would be carried in the spacecraft and their normal operation. The unsuited crew then conducted dry runs on procedures using training survival equipment and a mock-up trainer, including instruction on hatch operation, survival equipment stowage, and post-landing systems and activation. The second phase of training gave the crew egress procedures training while suited and the egress trainer in water. The second phase was divided into two parts; the first part of training was focused on having the crew upright the spacecraft using both normal and backup operations of the CMUS system and then conducting a Stable One egress, as shown in Figure 6.9. The second phase of the Phase II training, which was supervised by a medical team, trained the crew to conduct an egress operation in the event of an uprighting system failure, as shown in Figure 6.10. Because of the danger associated with the rapid buildup of CO_2 in the capsule, the crew would be forced to self-egress. But opening the docking hatch in a Stable Two orientation leads to the possibility of rapid crew compartment flooding once the hatch is opened. Following flat water egress testing, testing in the open water (Figure 6.11) was performed to familiarize the crew with a dynamic environment. The third phase of egress training instructed the crew as to their respective responsibilities in recovery operations during training exercises and actual missions. This third phase of training involved instruction by the project engineer on the egress and recovery procedures to be followed, including instruction on radio operations, use of the uprighting system, egress procedures, and helicopter pickup procedures.

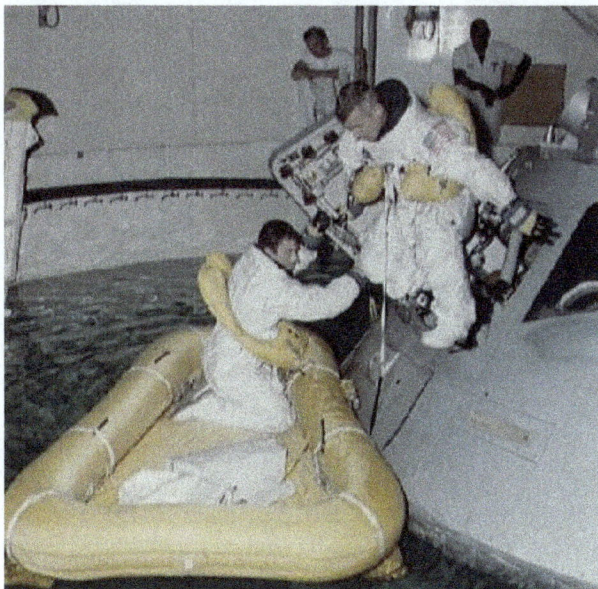

Figure 6.9 Apollo flat water egress testing in Stable One orientation (source: NASA)

Figure 6.10 Apollo Flat Water Stable Two egress testing (source: NASA)

Figure 6.11 Apollo open-water egress testing (source: NASA)

6.3.1   Apollo Egress Training

Initial spacecraft egress training was conducted in a training facility until crews were totally familiar with all procedures. Subsequent crew egress and evacuation exercises were conducted in the Vehicle Assembly Building at KSC or on the vehicle launch pad with assistance from ground support personnel and all evacuation equipment and facilities available to them, including elevators, slide wires, and blast rooms. Both flight crews and crew support personnel were then given first aid and fire training, consisting of four hours of lectures and demonstrations on 1) classification of fire types, 2) fire suppression, 3) extinguisher identification, selection, and use, 4) use of hoses, 5) respiratory problems in confined areas and breathing apparatus use, and 5) methods of protection from radiant heat. After first aid and fire training, crews were trained on the use of the slide wire system and actual runs were conducted with the crews in flight suits using the Portable Emergency Air Pack (PEAP), with and without helmets.

Total hours estimated for training were budgeted as follows:

Water Egress, Phase I …………………………………..	2 hours
Water Egress, Phase II …………………………........	4 hours
Water Egress, Phase III ……………………………….	6 hours
Spacecraft Egress Training …………………………..	2 hours
Launch Pad Egress Training..………………………..	4 hours
First Aid and Fire Training……………………………	4 hours

6.3.2   Apollo Post-Landing and Egress Procedure Checklists

Apollo egress procedures were divided into three sections: 1) stabilization after landing, 2) post-stabilization and ventilation, and 3) post-landing communications. Here, each crewmember is assigned specific tasks that are performed in rapid secession. Procedures consider both possible contingencies associated with low-power operations and Stable Two orientation and are detailed below:

Stabilization after Landing:
(note: CDR= Commander, CMP=Command Module Pilot, LMP=Lunar Module Pilot)

CDR	Set Emergency Locator System to 'AUTO' (Automatic)
LMP	Close both 'MAIN REL PYRO' (parachute release) circuit breakers
CDR	Release main parachutes
	Safe both Pyrotechnic and Logic Sequential Events Control Systems (SECS)
LMP	Open both 'Battery Relay Bus' circuit breakers
ALL	Take helmets off (if suited)
CDR	Close DIRECT O2 switch (if suited)

EGRESS PROCEDURES AND OPERATIONS

LMP	Turn off VHF AM B (center VHF radio) switch
CDR	Close Post-landing 'VENT' (ventilation system) circuit breaker
	Close all three 'FLOAT BAG' circuit breakers

If the CM is in Stable Two:
- Fill all three float bags until 2 min after CM flips upright, then disengage
- Turn off VHF AM A (VHF radio) switch while inverted
- Turn off VHP AM Beacon while inverted

If the CM is in Stable One, wait until after a 10 min cooling period, then:
- Fill all three Float Bags (takes approximately seven minutes)
- Turn off all three 'Float Bag' (CMUS activation) switches

Post Stabilization and Ventilation:

	Pull post-landing vent valve handle
	Remove post-landing vent exhaust cover
	Set post-landing vent to 'High' or 'Low'
	Turn on post-landing sea dye marker
	Release restraints (if suited)
LMP	Open 'MNA BAT BUS A' and both 'BAT C' circuit breakers
	Open 'MNB BAT BUS B' and both 'BAT C' circuit breakers
	Open 'FLT and PL BAT C' circuit breakers
CMP	Open 'PYRO A SEQ A' circuit breakers
	Open 'PYRO B SEQ B' circuit breakers
	CHECK DC voltmeter every hour to make sure it is above 27.5V. If not:

    Open 'FLT' and both post-landing 'BAT BUS A AND B' circuit breakers
    Close 'FLT' and Post-Landing 'VAT C' circuit breakers
    GO TO LOW POWER CHECKLIST
Unstow and install the post-landing ventilation duct
Deploy the grappling hook and line (if required)

Postlanding Communications:

- Set VHF Antenna to 'Receive'
- Set VHF Beacon to ON

If no contact; with recovery forces, then:
- Monitor VHF Beacon Transmission with Survival Transceiver

If the VHF Beacon is not operating, then
- Connect the Survival Transceiver to the antenna.
- Place VHF radio to 'Beacon Mode'

EGRESS PROCEDURES AND OPERATIONS

If the vehicle were to land in a state of low battery power, or to discharge to a low power state before recovery could be made, only essential systems would be activated. The following procedures would then be conducted:

LOW POWER PROCEDURES:

LMP	Turn VHF Beacon to 'Off'
	Set all three VHF Radios to 'Receive'
	Turn Floodlight to 'Off'
	Turn VHF AM B radio to 'Off'
CDR	Set VHF AM A radio to 'Receive Only'
CDR, LMP	Turn couch lights off
CMP	Plug Survival Radio into VHF beacon antenna cable
	Connect and turn radio on in beacon mode

If an emergency egress needs to be performed due to a post-landing emergency, procedures were established for both egress if the CM was in a Stable One orientation and a Stable Two orientation. Only side hatch egress methods and procedures were established. The docking hatch was not considered a viable means of egress.

EGRESS PROCEDURES: STABLE ONE

ALL	Disconnect umbilicals to suits (if suited)
	Put neck dams on (if suited)
CMP	Set center couch to the 270° position
CDR, CMP	Fold in armrests
CMP	Unstow survival rucksacks
LMP	Open side hatch
CDR	Turn off post-landing vent fan
CMP	Open all three 'BAT A, B, C, PWR ENT/PL' circuit breakers
CDR	Connect raft to capsule, if desired, with green lanyard
ALL	Connect raft white lanyards to water wings and inflate when egressing
CMP	Egress with liferaft
LMP	Put hardware kit out
LMP, CDR	Egress from CM

EGRESS PROCEDURES: STABLE II

LMP	Open all three 'CREW STA AUDIO' circuit breakers
ALL	Turn off all three 'POWER' switches
	Turn off all three 'SUIT POWER' switches
	Disconnect umbilicals (if suited)
	Release restraints (if suited)
	Set all three couch seat pans to the 170° position
	Put neck dams on (if suited)

CMP	Fold arm rests
	Remove survival kits from stowage
CDR	Connect liferaft mainline to CDR
CMP	Connect first white lanyard from liferaft to water wings
LMP	Connect second white lanyard from liferaft to water wings
CDR	Connect third white lanyard from liferaft to water wings
CMP	Open Pressure Equalization Valve
CMP, LMP	Remove and stow forward hatch
CMP	Exit feet first with rucksacks; when clear of CM inflate water wings and raft
LMP	Exit feet first with rucksacks; when clear of CM inflate water wings
CDR	Exit feet first with rucksacks; when clear of CM inflate water wings

6.4 Soyuz Egress Procedures

The Soyuz nominally lands on land, so egress should be less complicated than the American water landing vehicles. But unlike the American vehicles, Soyuz only has one egress path, through the overhead hatch shown in Figure 6.12. This 24.5 inch diameter hatch creates D-shaped translation path. The capsule provides only a 12 inch minimum distance between the hatch and the crew survival equipment when the hatch is open. No additional egress routes exist for emergency egress operations.

If the landing was at the primary or an alternate landing zone, egress should be fairly straightforward as the spacecraft will be on flat land, though if the parachute cutters fail, high winds could complicate the egress considerably. For a nominal landing, the Soyuz DM would land within 30km of the designated landing point and recovery teams in helicopters would approach the landing site soon after. Immediately after arrival the hatch is opened and an extraction stand is assembled to assist the Soyuz crewmembers to exit the spacecraft. Other helpers are responsible in cordoning off the area and gathering the landing parachutes.

But a Soyuz DM can and has landed in water – the errant Soyuz 23 that found itself in Lake Tengiz in -20° C conditions. Though few considerations were given this possibility before the events of Soyuz 23, self-egress in water is an operation that all subsequent flight crews have been trained in, even though the Soyuz Program still maintains water landings as low probability events. But an egress in water from a Soyuz DM is a very risky proposition as the symmetry of the Soyuz DM does not lend itself to stability in water and the module can fill with water and sink within one minute timeframe once the hatch is opened.

6.4.1 Nominal Soyuz Hatch Opening Procedures

The Soyuz DM has only one hatch – the one that connects to the OM while in orbit and is sealed before the OM is discarded prior to re-entry. Upon landing, the hatch may be opened by the crew or by a rescue team, though the close confines of the Soyuz DM (shown in Figure 6.12) make this a tricky operation. In the event of a contingency landing on foreign soil, basic instructions are provided in both Russian and English to assist in crew rescue. The instructions both request help, but also warn that the hatch may be explosively jettisoned. If an emergency is not present, rescue teams are advised to remain away from the DM until two antennas are automatically deployed approximately ten minutes after landing. Once the antennas are deployed, it is safe to approach the capsule. Rescue teams are then advised to avoid the cesium altimeter, which may emit a very low level of radiation to a radius of 15 ft from the bottom of the DM. To access the hatch at the top of the DM, a Hatch Opening Tool is available on the side of the DM, away from the cesium altimeter emission area. The tool can then be used to open the main hatch. The landing thrusters pose no hazard to rescue teams after they have been activated upon landing. Procedures are illustrated in Figure 6.13.

From inside, the hatch can be manually opened by rotating the lever that serves to retract six retaining rods. Following from discussions with NASA officials, once opened, it is believed that the Soyuz hatch could be closed "within a minute" in an emergency scenario. In events of delayed separation of the OM, such as those of Soyuz 5 and Soyuz TMA-11, aerothermal heating significantly damaged the hatch. As fortune would have it, the OM separated before a breach could occur, but a thermally warped hatch can create post-landing risk. In the case of Soyuz TMA-11, the hatch still functioned after landing but the capsules antenna was damaged, limiting communication, and the valve that equalizes pressure inside and outside the capsule was damaged.

Crew Survival Equipment   Hatch

Figure 6.12 US Astronaut Don Pettit showing the Soyuz egress path. Crew survival equipment is on left and the hatch is located to the right.

6.4.2 Nominal Soyuz Crew Extraction

The nominal crew egress procedure for the Soyuz assumes the crew to essentially be in a completely deconditioned where they could be treated as "unconscious trauma patients". After the hatch is opened (as detailed in Section 6.4.1), a medical assessment is immediately performed to determine the crewmembers health status based on external appearance and the results of questioning. If necessary, one medic may perform first aid with the crew still in the DM. At this point, visors are raised and gloves removed so that pulse readings could be obtained. Assuming the DM is on its side, the DM may then be rolled so that the crewmembers are in a "head up" position, facilitating egress. The crewmember located in the center seat is evacuated first, followed by the crewmember in the worst state of health.

If a crew member need be extracted in his seat liner, the hatch will need to be removed. Otherwise, egress begins with the removal of the soft portable emergency spares kits from the passageway to the hatch. The electrical connectors of cable communications and medical harness are disconnected with the connectors of the ventilation and gas mixture supply hoses of the spacesuits and the seat restraint system and footrests. After each crewmember is disconnected, they are extracted head-first, with the assisting rescue personnel taking the crewmember by the arms as the physician transfers the cosmonaut from the DM to a stretcher. If a crewmember has been injured, the spacesuit can be removed piece by piece using a knife or surgical scissors.

6.4.3 Cosmonaut Egress and Survival Training

Wilderness survival training is provided to all cosmonauts to fly on Soyuz missions. The relatively high inclined orbits of Soyuz missions led the Russian Space Agency to emphasize winter survival training, which was initially conducted in Vorkuta, in the Russian Arctic since the 1980's. It is now being conducted near Star City. Here, flight crews would learn to survive for up to three days, learning skills such as how to build a snow shelter as a means to improve heat conservation. Crews were provided with parachute material as well as equipment that could be expected to be provided in survival conducted in Uzbekistan, where flight crews would learn to use parachute material to make a tent and hammock, start a fire, and use radios, a pistol, and conserve food and water. Mountain training is also provided, as cosmonauts have recently trained with NASA astronauts in the mountains of Utah, scaling a 4,500m mountain and hiking 60km. Swamp survival skills are also covered.

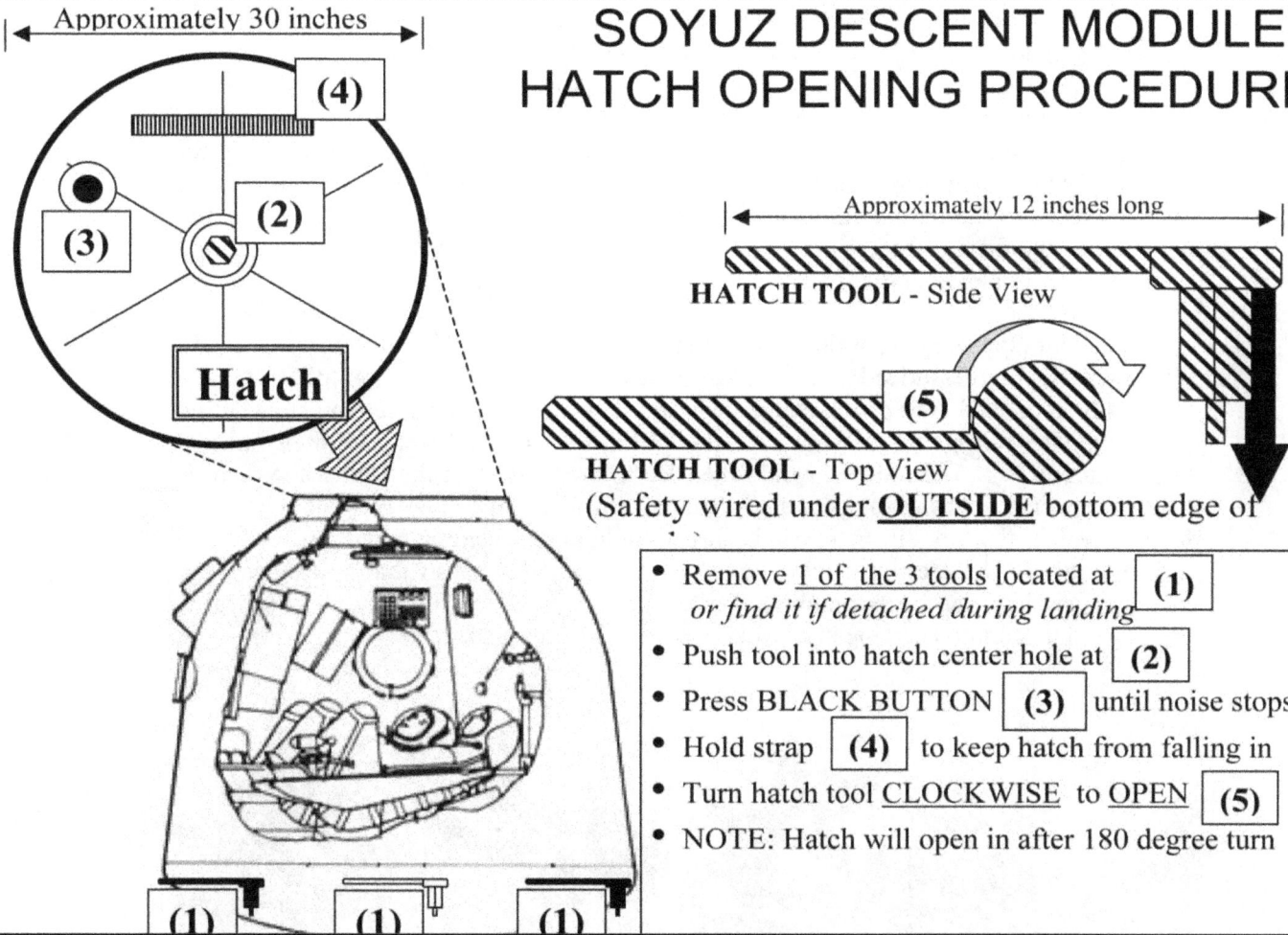

SOYUZ DESCENT MODULE HATCH OPENING PROCEDURES

Approximately 30 inches

(4)

(3)

(2)

Hatch

Approximately 12 inches long

HATCH TOOL - Side View

(5)

HATCH TOOL - Top View
(Safety wired under **OUTSIDE** bottom edge of

- Remove 1 of the 3 tools located at (1) *or find it if detached during landing*
- Push tool into hatch center hole at (2)
- Press BLACK BUTTON (3) until noise stops
- Hold strap (4) to keep hatch from falling in
- Turn hatch tool CLOCKWISE to OPEN (5)
- NOTE: Hatch will open in after 180 degree turn

(1)   (1)   (1)

Figure 6.13 Soyuz Descent Module Hatch Opening Procedures (source: US Air Force Rescue Coordination Center)

147

Sea survival training is an integral part of a cosmonauts training routine as well, even though the Soyuz DM is designed to land on land. Sea survival training is conducted so as to acquaint the astronauts with the techniques and procedures used for surviving in the sea until rescue assistance arrives. A water landing is simulated by dropping a boilerplate capsule off of a ship in the Black Sea. Here, the crews learn to egress the DM, ingress a raft, and don sea survival outfits. Crews also learned sea survival techniques. It is interesting to note that Soyuz crews were provided with TZK thermal protection suits and "Forel" immersion suits in a vehicle designed to land on land whereas Apollo (and presumably Orion) crews that landed in water nominally were not equipped with immersion suits and would have cold water protection and buoyancy as a requirement for the space suits. Arguably, the impact of the near LOC event of Soyuz 23 drove the program to adopt these precautions to protect against similar events in the future. It is to note that cosmonaut Sergei Yuriyevich Vozovikov drowned on 11 July 1993 during water recovery training in the Black Sea, near Rayon Anapa, Russia

As part of the Soyuz egress training, cosmonaut trainees would learn to ingress the DM with their Sokol pressurized space suits donned. Once inside the cramped DM, the trainees then donned their cold weather gear and their waterproof immersion suits before egressing one at a time by jumping from the DM hatch into the water with their survival kits (Figure 6.14). The trainees would then learn to egress from the DM when there is insufficient time to change suits and while equipped with only emergency flotation bags for survival in the sea. In the DM, in direct sunlight, temperatures can quickly cause hyperthermia in the crew and crews must learn to collaborate so that suit egress and ingress operations can be as timely and effective as possible. Then once in the water, crews need to learn to keep together while waiting to be rescued.

Figure 6.14 Water Egress training from Soyuz DM using immersion suits (source: Roskosmos)

The Soyuz 23 capsule landed in Lake Tengiz in -20 deg C conditions in a snowstorm. The wet parachute filled and dragged the capsule below the surface, cooling the capsule. Heating systems had to be shut down to conserve battery power. The crew hung upside down, trapped inside the spacecraft because the exit hatch is located on the top of the capsule. Further, no fresh air could enter the spacecraft because the valve that allows the air in was now under water. The sinking risk was considered too great to risk opening the hatch so the crew was forced to rely on a set of dying batteries to keep the air regeneration equipment running long enough for rescuers to find them. Cosmonauts Vlaery Rozhdestvensky, and Vyacheslav Zudov, like the crew of Apollo 13, had devised a strategy to conserve the precious and life sustaining electricity for the air rebreathing equipment.

When the rescue teams arrived, amphibious vehicles attempted to recover the spacecraft but could not reach it. Many hours later, swimmers managed to attach a cable to a helicopter so that it could be dragged many kilometers across the lake, and only in the morning was the crew able to be extracted from the DM. The recovery crews were surprised to find the crew alive, though both cosmonauts were rendered unconscious from hypothermia and oxygen depletion.

The cosmonauts acted according to standard procedure by not jettisoning the parachute after landing, but the situation they were in was unprecedented. No contingency procedures existed for a scenario where the DM might land upon a frozen lake! Consequently, the parachute sank and pulled the DM upside down, placing the hatch and pressure relief valve, which had opened as the DM fell through an altitude of five kilometers, under water. After the flight, neither cosmonaut suffered any lasting effects, but neither also flew another mission. Soyuz 23 demonstrated that the Soyuz rescue and recovery system functioned adequately and was responsive to contingencies. The Soviet news agency later commented on the flight. It reported: 'At all stages of the flight and after landing, the crew acted in a confident way, effectively discharging their duties.

In Cosmonaut Zudov's Own Words:

"...We had no possibility to get out of the capsule, as we did it before and as we were taught. If we open the hatch, and it was under water, the stream will come onto the cosmonauts, and there would be no chance to get out. Besides – very low

temperature on water. Temperature on Lake Tengiz was -22°C.. To stay in such situation in the spacesuits and do nothing – we would be frozen and die. That's why first we had to do – to get out of spacesuits and to free ourselves of them. We spent and hour and a half to get out, even used knives to cut them. Then we managed to wear our plain sport wear. Started to save electric power. Switched off all sources inside the capsule – left only radio beacon and radio station which we could use in due time – but it failed soon..."

Case Study: Soyuz 23 (continued)

Cosmonaut Rozhdestvensky Describes:

"...You could feel CO_2 without any instrumentation, just feel it. When I felt that it at the stage – which we could loose consciousness – then I switched on the regeneration unit. Then, when the mind started to clear, and blue halo we were starting to see disappeared in the eyes – I switched it off. And so on – all the night. Helicopters could do nothing there. In the frozen blowing mist, which covered all the lake – helicopters could not go down. The crew was trapped like in a can. Divers also could do nothing in that salted quagmire. All efforts to put a plate under the capsule or turn it around – failed. That's why, finally, there was an order to hook the capsule – over the parachute – and drag it to the shore by helicopter. They did it and dragged the spacecraft for around five to seven kilometers. When I got to see our photos and how they dragged the capsule – then I really was frightened – the only time in my life I was really frightened."

Figure 6.15 Soyuz 23 landing in Lake Tengiz

In addition to Soyuz 23:

In addition to Soyuz 23, the Soyuz 10 landing of April 24, 1971 was also headed directly to a lake, though just before it would have splashed down, a gust had pushed it onward towards a shore, where it landed a mere 44 meters from the water's edge.

150

6.5 Space Shuttle Egress Procedures

The Space Shuttle developed egress procedures, though initially they all assumed a pad egress scenario or a landing at or near a suitable landing strip. Only after the Challenger disaster was a bail-out "Mode VIII" egress incorporated. Protective 'ACES' suits weighed 32 kilograms (70 pounds) and were used to protect the crew from ingress into the orbiter on the pad until safely in orbit. These suits protected the occupant from most hazardous environments and included an emergency oxygen system, a parachute system that included an automatically actuated pilot, drogue and main parachute; a seawater activation release system; flotation devices; a life raft and survival equipment. A listing of the egress modes used during the Space Shuttle Program are provided in Table 6.6.1.

6.5.1 Mode I: Unassisted Pad Egress and Escape

A Mode I egress is an unassisted pad egress involving an escape to a secure area. A Mode I egress will be called for if the emergency situation is too critical to dispatch fire/rescue personnel to the pad and the closeout crew and the Astronaut Support Persons (ASPs) have left the launch platform. After a Mode I egress is called, the orbiter crew must do the following without assistance:

1. Unstrap themselves and egress their seats.
2. Open the side hatch.
3. Proceed to the emergency slidewire baskets (Figure 4.)
4. Descend in the baskets.
5. Take refuge in the concrete bunker and, if directed, evacuate in an M-113 armored personnel carrier (Figure 4.).

The emergency escape system used for pad emergency egress scenarios includes seven baskets suspended from seven slidewires that extend from the fixed service structure to a landing zone 1,200 feet to the west. The seven slidewire baskets are enclosed in fire-resistant material and can each can accommodate four persons with suits on. Once in the bunker, the crew will follow directions whether to either stay in bunker until rescuers arrive or evacuate in M-113 armored personnel carrier. The bunker has breathing air, water, phone line communications, and medical supplies.

Mission Phase	Type		Mode	Description
Prelaunch	Pad egress and escape	√	I	Orbiter crew unassisted.
			II	Orbiter crew assisted by closeout crew.
			III	Orbiter crew assisted by fire rescue crew.
			IV	Orbiter crew and closeout crew assisted by fire rescue crew.
Postlanding	Orbiter egress and escape	√	V	Orbiter crew unassisted. Egress may be any of three types: a. Hatch on b. Hatch jettison c. Window 8
			VI	Landing on or near runway. Orbiter crew assisted by pre-positioned convoy crew.
			VII	Landing in a remote area. Orbiter crew assisted by rescue and medical personnel arriving by helicopter.
Ascent or entry (for controlled, gliding flight)	Bailout	√	VIII	Bailout from orbiter in following circumstances: a. During ascent or entry b. Over land or water

Table 6.6.1 Space Shuttle Emergency Egress Modes (credit: United Space Alliance)

Step	Crew	Action
1		**Prep for egress**
	All	Upon call by CDR or NTD for emergency egress: a. Check that neck dam tabs are released. b. Close and lock helmet visor. c. Turn on orbiter oxygen to conserve Emergency Oxygen System (EOS) oxygen d. Activate EOS by pulling green apple. e. Remove kneeboards and discard. f. Release quick disconnect for cooling. g. Release seat restraints by turning rotary buckle. h. Release parachute Frost fittings and ejector snaps. i. Release quick disconnects for comm. j. Release quick disconnects for Orbiter oxygen. k. Egress seat.
2		**Side hatch opening**
	MS3	a. Egress seat 5 carefully, using handhold strap on forward locker. b. Climb down to egress platform; proceed to side hatch c. Rotate hatch handle to the VENT position, then rotate to the OPEN position. d. Open side hatch.
3		**Evacuate orbiter**
	All	Crawl out side hatch headfirst.
4		**Proceed to baskets**
	All	Follow marked route through the white room and across the Orbiter Access Arm and the Fixed Service Structure to the baskets. Alternate route is used if primary route is blocked.
		Firex system activation
		Activate the Firex fire extinguishing system if the system has not been activated by the Launch Control Center.
5		**Descend in slidewire baskets**
	All	Descend as follows: a. Two to three persons enter each basket. b. Using the manually operated guillotine-type device, the crewmember nearest the guillotine in each group severs the rope securing the basket. c. All ride baskets to landing zone (takes 22 sec). d. After basket stops in landing zone, a crewmember in the basket pulls a D-ring to release the south side of the basket for easier egress. e. Exit basket and go directly to the bunker being sure to stay on cement walkways.

Table 6.2 Mode I Egress Procedures (credit: United Space Alliance)

6.5.2   Modes II, III, and IV: Assisted Pad Egress and Escape

In Modes II, III, and IV, the three assisted pad egress and escape modes, support personnel assist the flight crew in egressing from the orbiter and escaping to a safe place. Differences among these modes are discussed below. Egress following shutdown of the Space Shuttle Main Engine (SSME) also is discussed.

At call for mode egress, crewmembers immediately perform the following steps:

1. Close and lock visor.
2. Turn on orbiter O_2.
3. Activate EOS.
4. Pull quick disconnect for cooling.
5. Pull quick disconnect for comm.
6. Pull quick disconnect for orbiter O_2.

The egress will then proceed as directed by the rescue team leader, which will depend on the type of egress operation being conducted. As listed in Table 1.1, the closeout team leader is responsible for the execution of a Mode II egress while the rescue team leader is responsible for the successful execution of a Mode III or a Mode IV egress. Procedures are then conducted as are listed in Table, after which the crew with the rescue team will descend from the gantry using the emergency slidewire baskets, enter the bunker, and follow the NASA Test Director (NTD) as to whether to remain the bunker or to evacuate using the M-113 armored personnel carrier. In all cases, the triage doctor is responsible for triage and medical evacuation decisions are made by the flight doctor.

Mode	Fixed Service Structure	Bunker	Triage	Evacuation
I	CDR	NTD	Triage Doctor	Flight Doctor
II	Closeout Leader	NTD	Triage Doctor	Flight Doctor
III, IV	Rescue Leader	NTD	Triage Doctor	Flight Doctor

Table 6.3 Responsible persons in phases of an emergency pad egress scenario (credit: United Space Alliance)

In a **Mode II** egress, the crewmembers are assisted by the closeout crew. The procedures are essentially similar to that of a Mode I egress except the closeout crew assists the crewmembers out of their seats (if necessary), ensures that oxygen is being supplied to crewmembers without helmets, and then assists in the evacuation to the slidewire baskets.

A **Mode III** egress assumes that some crewmembers may be unable to egress on their own. If some crewmembers can egress on their own, they open the side hatch and assist the other crewmembers to the slidewire system. If the crew is completely incapacitated, then the closeout crew 1) opens the side hatch and enters the Orbiter, 2) checks that each crewmember's helmet and visor is closed and that the EOS bottles are activated, and 3) assists the crew egress either down the elevator (time permitting) or via the slidewire system.

A **Mode IV** egress is essentially similar to a Mode III egress except this egress mode assumes the possibility that the side hatch may still be open and that there might be a combination of both flight crew and closeout crew in need of assistance.

6.5.3   Mode V: Unassisted Post-Landing Egress

A Mode V egress operation may be conducted at any time in the post-landing phase of a Shuttle Mission. This egress, performed without assistance from convoy crew personnel, may be conducted in one of three variations:

> 1.   A hatch-jettison Mode V egress can be made when there is imminent danger to the orbiter or the crew; e.g., fire in the crew module.

> 2.   A hatch on Mode V egress can be made when an egress has to be expedited because of a hazardous condition or when the landing occurs at a landing site that has no convoy crew support.

> 3.   A Window Eight Mode V egress may be made when the side hatch egress path cannot be used.

The primary egress path is through the side hatch, descending to the ground via the Emergency Egress Slide, as shown in Figure 6.16, or via descent control devices in the event of a slide failure. The Emergency Egress Slide inflates with a self-contained supply of air, allowing a safe exit to the ground within one minute of the hatch opening. If the side hatch cannot be opened or jettisoned, or if a hazardous environment exists outside of the side hatch, an egress may be performed through the Window Eight escape panel. Window Eight egress operations always rely on the 'Sky Genie' descent device. If communication is lost with the ground support crew, the crew can communicate via the PRC-112 survival radio or by using light signals.

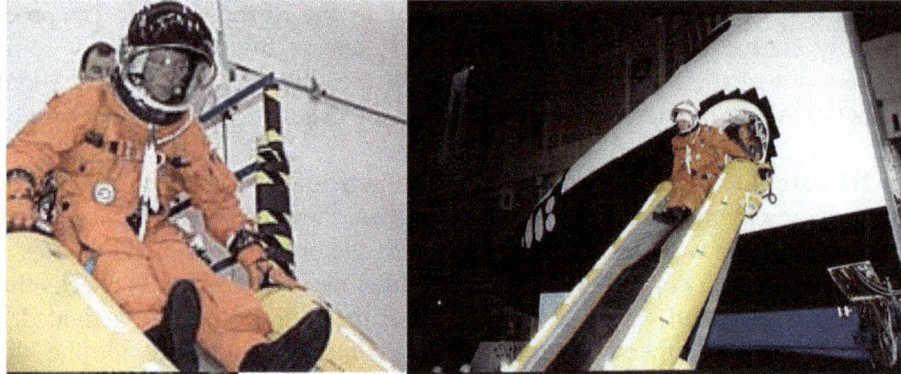

Figure 6.16 Side hatch egress using the escape slide (source: aerospaceweb.com)

Using the Sky Genie

To egress using the Sky Genie, the nylon line upon which the crew members will descend is secured at an anchor point near the egress port (presumably window eight) and crew members, in turn, secure themselves to the Sky Genie (Figure 6.17). Once fastened, the crew member must maintain tension on the line below the Sky Genie and begin a descent by gradually easing tension. Control of the free end of the line is essential.

Assuming the crew members is capable of controlling their own descent, the crew member may regulate the rate of descent by applying and easing downward tension on the line feed. If an unconscious or injured crew member needs to be descended, this may be done from above with the assistance of another crew member. In this case, the person controlling the descent from above firmly holds the unsecured line until the incapacitated crew member is ready for descent and then gradually eases tension on the unsecured line.

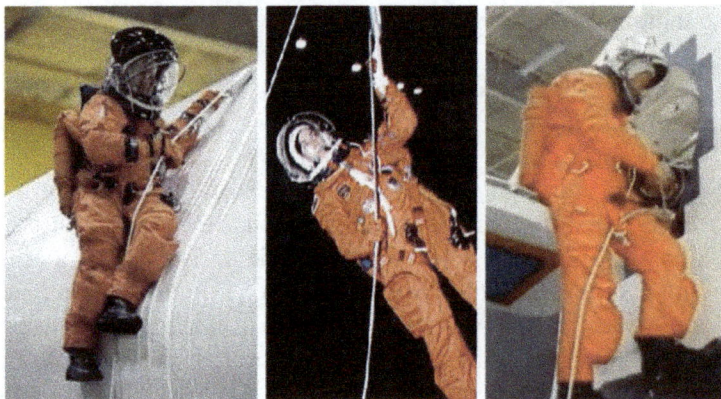

Figure 6.17 Crewmember egressing via Window Eight using the Sky Genie (source: aerospaceweb.com)

EGRESS PROCEDURES AND OPERATIONS

Jettisoning the Side Hatch

If there is need to remove the hatch and the crew is unable to open the hatch, or if the required egress is extremely time critical, the hatch can be explosively jettisoned. To jettison the hatch using explosives, MS3 will open the pyrotechnic control box by 1) removing the safety pin and squeezing the latch knobs together to pull the cover down, 2) removing the hatch jettison T-handle pip pin and white cloth cover, and 3) squeezing and pulling the ribbed hatch jettison T-handle so that the handle will come free from the base and the hatch may be jettisoned.

Step	Crewmember	Procedure
1		**Call for Mode V Egress**
	CDR	Call for Mode V Egress
2		**Seat Egress**
	All	At CDR call for Mode V egress, egress seat per cue card
3		**Slide Deploy**
	MS3	a. Unlock hatch handle and rotate handle counterclockwise to VENT position per hatch decal. b. Pull pin to release slide cover handle. c. Lift handle and remove slide cover. d. Lift girt bracket with the slide pack; lock bracket in place and rotate slide assembly ~270° into side hatch tunnel. e. Remove aft, then forward hinge pip pins. f. Lift and rotate entire slide pack back 90° and insert girt bracket into slide attach points (decals) on the hatch. g. Make sure slide is fully locked in place. h. Rotate hatch handle to the UNLATCHED position and push hatch open. Crawl out onto side hatch and push the slide off the outboard edge of the hatch. j. Pull the inflation lanyard all the way out without releasing grip.
4		**Hatch egress, slide descent**

	All	All crewmembers egress through the side hatch as follows:
		a. Crawl onto the side hatch and descend slide feet first, making sure to:
		    1. Keep legs and feet slightly apart and toes pointed up.
		    2. Keep hands at side of body to aid balance.
		    3. Lean back with weight off heels.
		    4. Stand and move away upon reaching bottom of slide.
		b. Upon reaching bottom of slide:
		    1. Use forward momentum to assist in standing.
		    2. Move to one side.
		    3. Assist next person.

5		**Escape to safe area**
	All	Move crosswind to edge of runway, and then move upwind

Table 6.4 Space Shuttle Mode V Egress Procedures (credit: United Space Alliance)

6.5.4   Mode VI and VII: Assisted Post-landing Egress

A **Mode VI** or **VII** post-landing egress is similar to a **Mode V** egress, except that rescue personnel are available to assist the crew both in egressing from the orbiter and in escaping to a safe place. A **Mode VI egress** occurs on or near a runway that is readily accessible to pre-positioned convoy ground personnel. A Mode VII egress is triggered when the Orbiter lands or crashes off the runway within 25 n. mi. of the intended landing site. Once a **Mode VII** abort is declared, rescue and medical personnel are transported to the crash site by helicopter.

6.5.5   Mode-VIII: 'Bail-Out' Egress

A **Mode VIII** emergency egress is conducted from controlled gliding flight at an altitude of 30,000 ft or less when a safe landing is unlikely. Bailout could occur during:

a.   any ascent abort (RTLS, TAL or AOA)
b.   an emergency de-orbit that must be made regardless of landing site.
c.   any failure precluding a normal landing on a runway.

During a bailout, the pilot would attempt to bring the Orbiter onto a level flight path at a relatively low altitude and speed. The Orbiter's flight controls also included an autopilot mode engaged by the pilot or commander that would maintain an automatic stable flight while the crew escaped. Once safe flight conditions had been achieved, the crew would have opened the vehicle's main entry hatch to extend an escape pole through the hatch and out into the airstream. One after another, each crewmember would have attached a hook assembly from the escape pole to his or her parachute harness and jumped out of the Orbiter through the hatch. The crewmember would have slid down the pole and off the end along a path that (hopefully) would have carried each astronaut below the left wing of the Orbiter. Once safely away from the vehicle, each crewmember's parachute would have opened for a safe descent to the surface below. Once the crew had egressed, the Orbiter would have crashed.

At an altitude of about 9,150 meters (30,000 feet), astronauts would pull a handle that turns on the depressurization valve in the crew compartment bulkhead. This equalizes the cabin pressure and outside air before the side hatch is released. At 7,620 meters (25,000 feet), the hatch is jettisoned and two telescoping sections of the aluminum and steel escape pole deploy through the hatch. Each crew member hooks a Kevlar® strap onto the pole, jumps out the hatch opening, slides down the 3.1-meter (10-foot) pole and goes into a freefall until the parachute opens to ease the journey to the ground. It takes approximately 90 seconds for a crew of eight to bail out of the Space Shuttle, and by that time, the vehicle is at 3,050 meters (10,000 feet) altitude.

The **Mode VIII** abort option assumes the Orbiter had already separated from its Solid Rocket Boosters and External Tank and was in controlled gliding flight. During a bailout, the pilot would attempt to bring the Orbiter onto a level flight path at a relatively low altitude and speed. The Orbiter's flight controls also included an autopilot mode engaged by the pilot or commander that would maintain an automatic stable flight while the crew escaped. Once safe flight conditions had been achieved, the crew would open the vehicle's main entry hatch to extend an escape pole through the hatch and out into the airstream pas the left wing. One after another, each crewmember would attach a hook assembly from the escape pole to his or her parachute harness and jumped out of the Orbiter through the hatch. Each crewmember would then descend by parachute and the orbiter would be lost.

Altitude (ft)	Step	Crewmember	Action
50,000	1		Call for bailout

		CDR	If range to go exceeds ~ 55 n. mi. at 50,000 ft:
			    a. Call for bailout.
			    b. Place orbiter in minimum sink rate attitude.
			    c. Engage autopilot.

		All	At CDR call for bailout:
			    a. Lower and lock helmet visor.
			    b. Suit O2: ON
			    c. Pull green apple.

40,000	2		Cabin Venting
		CDR	Request MS3 to vent cabin.
		MS3	At CDR request to vent cabin:

			a. Vent cabin as follows:
			   1. Open pyro box.
			   2. Pull safing pin on cabin vent T-handle.
			   3. Squeeze and pull smooth pyro vent T-handle.
			b. During cabin vent, monitor altimeter on middeck locker and communicate reading to CDR.

		CDR	During cabin vent, monitor altimeter reading and communicate reading to MS3.
~31,000			Cabin pressure equalization
	3		Seat Egress Preparation
		All	During cabin vent: a. Remove kneeboards. b. Break cooling and any biomed connections. c. Release seat restraints by turning rotary buckle. d. Remove cover to parachute D-ring on right riser and extend D-ring.
30,000	4		Side Hatch Jettison
		CDR	When pressure equalized, request MS3 to jettison side hatch.
		MS3	At CDR request to jettison side hatch: a. Pull safing pin on hatch jettison T-handle. b. Squeeze and pull ribbed hatch jettison T-handle (forward handle).
		All	c. If bailout after 2 days in space, pull *g*-suit clip. d. Disconnect comm and orbiter O2.
~30,000	5		Escape Pole Deploy

		All	a. Egress seat and fold down seat back.
			b. Follow decal to deploy escape pole.
			If pole does not deploy fully, deploy pole manually:
			   1. Pull manual backup pip pin.
			   2. Pump ratchet handle in direction of pole (power stroke is upward) until green stripe visible.

~30,000	6		Bailout
		All	At side hatch: 1. Attach D-ring on parachute riser to snap hook on lanyard. 2. Release lanyard from magazine. 3. Kneel into hatch. 4. Remain tucked and forcibly roll out of hatch.

Table 6.5 Mode VIII Egress Procedures (credit: United Space Alliance)

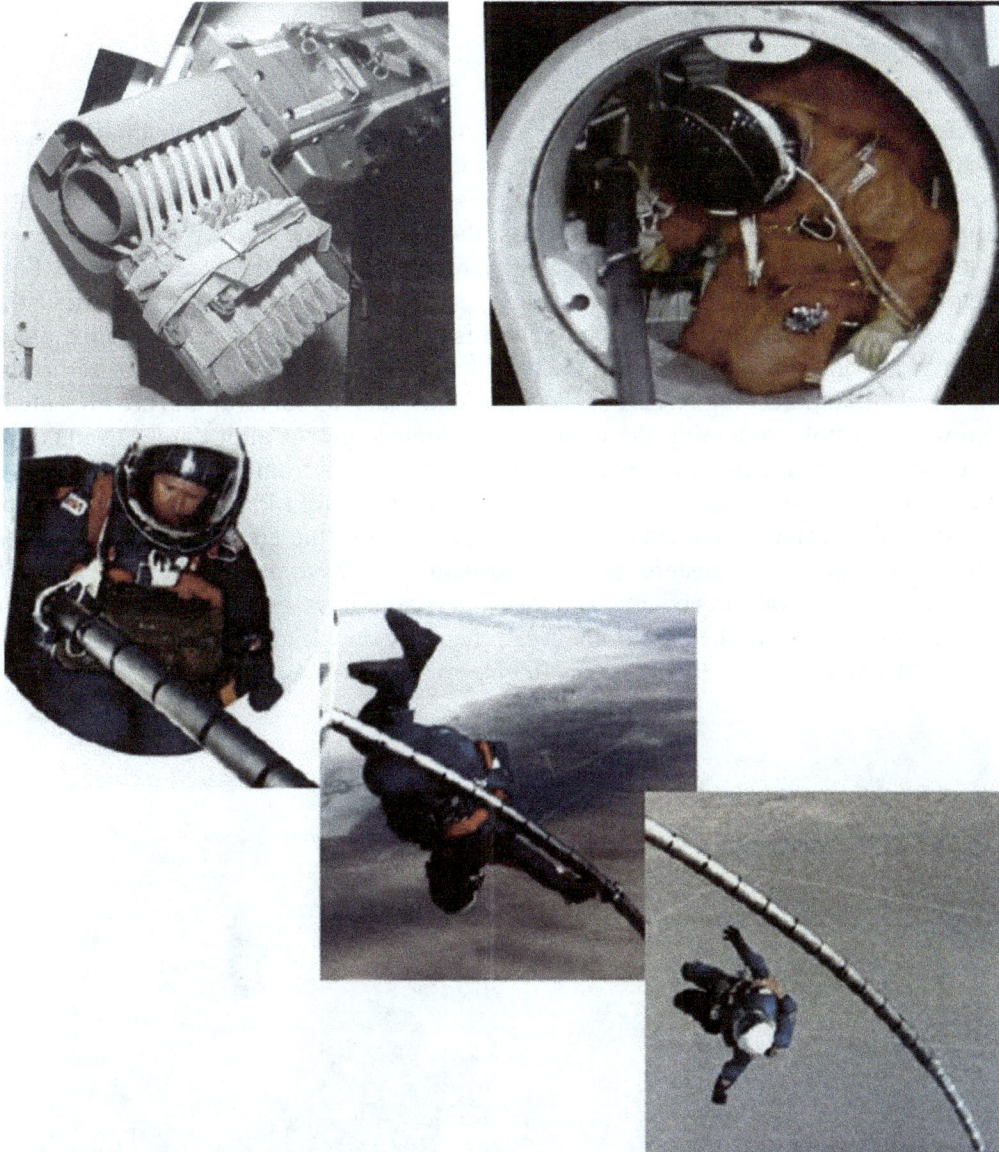

Figure 6.18 Space Shuttle emergency 'escape' pole (source: aerospaceweb.com)

6.6 Constellation Egress Procedures

The Orion capsule had similar challenges to the Apollo capsule, though egress procedures are complicated as the four-person crew was stacked so that the feet of two crewmembers would be secured above the heads of the other two. The added congestion complicated egress operations. Nevertheless, the key to successful astronaut rescue was still to remain in the CM and await aided crew egress, but if necessary, the Flight Director or Crew Commander will direct an expedited emergency unassisted evacuation, due to a severe post-landing contingency. In this event, the primary egress path is out the side hatch and then into the raft; the docking hatch is the alternate path if side hatch is unusable. Unlike preceding NASA spacecraft, items such as survival kits are allocated to the crew as a whole, assuming the crew must remain together after egress to maximize chances of survival and to ensure rescue. Egress procedures from the Orion capsule were refined through the Post-landing Orion Recovery Test (PORT) Phase II. Several tenants were established prior: 1) personal flotation is provided to the crew (suited or unsuited), 2) the crew enters an austere survival situation with minimal shelter capability; knowledge of survival procedures and use survival kit is essential, and 3) the suit provides protection from the environment but the crew decides whether to depart the CM suited or unsuited based on their best judgment.

Figure 6.19 Egress testing out of low-fidelity Orion mock-up (source: NASA documents)

6.6.1   Constellation Post-Landing Orion Recovery Test Phase Two (PORT II)

The PORT series of tests were a series of three tests designed to test the characteristics of the Orion spacecraft after splashdown.   The first phase test, conducted in 2009, determined the ability of rescue and recovery teams to affect a rescue of the crew and a recovery of the capsule.   The second phase, PORT Phase Two, would determine the ability, method, and time required for the crew to perform an assisted and an unassisted egress in varying sea conditions via each of the spacecraft's two hatches: the side hatch and the docking hatch.   The test would also determine the ability of the crew to safely ingress a liferaft from each of the hatches and test methods of attaching the raft to the spacecraft.   Design of the vehicle would be influenced by this test as information regarding egress aid (handle) positioning, raft and survival kit positioning, and descent aids should the crew need to egress out of the docking hatch.   Further, lateral seat supports and umbilical attachments were tested to determine their negative effect upon egress time.   Finally, suits were to be doffed in a test to determine the effectiveness of the crew to perform this complicated task while being influenced by the cramped conditions and oscillating environment of the Orion capsule after splashdown.   Initial 'expert opinions' would then be gathered from prior flight crew and aerospace medical experts on the effectiveness of injured or incapacitated crewmembers.

Figure 6.20 PORT Egress Testing (left).   Author testing Orion mockup (right) (source: NASA PORT study summary)

6.6.2   Orion Post-Landing Unassisted Egress Procedures

The procedures provided in Appendix A were developed for the Post-Landing Orion Recovery Test Phase Two (PORT II) to validate a methodology to conduct effective unassisted egress operations from the Orion spacecraft. Nominal egress procedures are provided, followed by procedures in the event that an egress must be conducted through the docking hatch or the CM must be inverted through use of the CMUS system. The procedures of Appendix A are performed once an emergency egress has been declared and the initial procedures have been performed. The stable-one orientation can be achieved either by landing or through a successful activation of the CMUS after a landing in a Stable Two orientation. Assuming that the air suit loop is exhausted or otherwise unavailable, the egress procedure must be performed using the EBS system.

The initial procedures are assumed to be performed shortly after landing. If a hazardous environment is the cause of the unassisted egress, then it would certainly not be advisable for any crew member to open their suit visor, as once the first crew member breaks the mask seal by raising their visor, all crew members will be exposed to the environment within the capsule. Assuming that a hazardous environment does exist, the commander will verify landing and that the internal air is free of toxic contaminants (e.g., hydrazine, ammonia, smoke), and attempt to contact the recovery forces. All crewmembers may then open their suit visors and remove their gloves.

Unassisted side hatch egress procedures were developed if there were to be a need for an unassisted emergency egress and sea conditions are benign enough so that there would not be a sinking risk if the side hatch were opened. In addition, a docking hatch egress procedure was developed if the vehicle is in a Stable One orientation, there is need for an unassisted emergency egress, and it is deemed that opening the side hatch would be too risky due to the pitching of the vehicle and the associated risk of a rapid sinking situation. Furthermore, if the Orion capsule lands in a stable-two orientation and the Command Module Up-righting System (CMUS) fails, the crew is left in a vulnerable situation. As the design of the spacecraft slightly favors a stable-one orientation, it is possible that it may orient itself that way as a result of sea motion, but sea conditions that agitated may make an unassisted an egress extremely risky proposition, but then again, so is remaining in the spacecraft with an uncontrollable hazardous environment. All of these procedures are provided in Appendix A.

If the CMUS system fails, the crew will be suspended by their seats in an inverted position. It is important that the crew extract themselves from the suspended position, as crews so suspended have a limited period of consciousness, and it may take up to three minutes to verify that the CMUS has, indeed, failed. The following set of procedures assumes that the egress initiator has occurred within the first three minutes following landing, the crew remains suspended in their seats, and it has been determined that the CMUS has failed.

EGRESS PROCEDURES AND OPERATIONS

6.6.3   Orion Non-Emergency Self-Egress in Excessive Sea Conditions

There is a possibility that a launch could be conducted with an acceptable degree of risk that if an ascent abort were triggered at a particular time, the spacecraft could land in sea conditions in excess of what the recovery team could safely conduct a rescue operation in. As a rescue operation can not be safely conducted in an environment that has not been validated through previous testing, an operational threshold is established.

The Orion spacecraft is designed to effectively target away from the North Atlantic between St. John's, Newfoundland and Shannon, Ireland. We can safely assume that there negligible risk of landing within this zone, but there is significant probability of landing in or about the targeted zones around these two regions, and sea conditions may be greater than a rescue team can safely conduct a rescue in and winds may be greater than PJ teams can safely deploy in. This is especially a concern in wintertime. If, however, the Program accepted only zero probability of landing in such conditions, launch windows would be exceedingly rare and approach zero in wintertime. By accepting a slight risk of landing in an excessive environmental condition, we reap huge gains in the overall operability of the Program.

So the question must be addressed is: What do we do if an ascent abort causes the spacecraft to land in excessive environmental conditions?

As this condition is valid only on ascent, it is safe to assume that no crew member will be deconditioned, but due to possibility of landing on a wave face in excess of Orion design specifications, the injury risk to the crew will be elevated. As PJs would be unable to deploy or assist, the crew would need to conduct their own egress into the water and then be extracted by helicopter. To avoid the risk of a rapid flooding and sinking situation, a docking hatch egress is assumed, unless injuries or a time constraint (e.g., from a post-landing failure) make this impossible, in which case side hatch egress may be the only option.

The Orion spacecraft was assumed to be in a Stable One orientation with a helicopter and an AMVER registered vessel in close proximity. The greatest risks will be incurred during the actual process of egress. The following risk drivers were considered most significant during egress: 1) a crewmember is injured from a fall inside the CM, 2) a crewmember is injured from a fall outside of the CM, 3) a crewmember is hit by the pitching CM while in the water, 4) the CM flips to Stable Two with the hatch open, and 5) if a side hatch egress is necessary, the CM floods and sinks before the crew can egress. After egress, the following risk drivers were considered most significant: 1) a crewmember becomes trapped under CM, 2) the crew cannot ingress the raft, 3) a Life Preserver Unit fails, 4) the effects of hypothermia or dehydration set in before rescue, and 5) the crew encounters hazardous sealife before rescue.

6.7 Sierra Nevada's Dream Chaser

Certain trajectories may expose this spacecraft to periods of time where a suitable landing zone may be targeted following an ascent abort. Its predecessor, the HL-20, had parachutes that could be used to orient the spacecraft for a tail-down water landing, so it is likely that Sierra Nevada is investigating this method to safeguard times on the ascent profile when a landing strip could not be targeted. Further, the HL-20 is based on the Russian BOR-4 spacecraft, which was orbited several times by the Soviets starting in 1982 and recovered in the Indian Ocean and Black Sea. During recovery operations of the space plane in the Indian Ocean, an Australian P-3 Orion aircraft obtained photographs of the vehicle both floating in the water and being hauled aboard the recovery ship.

The design of the HL-20 concept had taken into account crew safety and survivability for various abort modes. The interior layout with a ladder and hatch arrangement was designed to permit rapid egress of passengers and crew for emergencies on the launch pad. For on-the-pad emergencies or during launch where time was a critical element (launch vehicle fire or explosion), the HL-20 would be equipped with emergency escape rockets that would rapidly thrust the PLS away from the booster. The method was similar to that used during the Apollo Project. Once at a safe distance, a cluster of three emergency parachutes would open to lower the vehicle to a safe ocean landing. Inflatable flotation devices ensured that it rode high in the water, with at least one of two hatches available for crew emergency egress.

7   Assessing Probabilities and Effects of Injuries and Deconditioning

As much of the preceding chapters have elaborated upon, the landing of a spacecraft, especially a capsule-based spacecraft, is an operation that is highly variable on the terrain and surrounding environments. Accidents can and do happen; spacecraft impact the ground in ways that can impart excessive impact velocity, or excessive side forces, to their occupants. Further, the human occupants themselves are highly variable; some will respond better to such impacts than others either by virtue of their physiology, their state of deconditioning, or by the manner in which they are secured to and protected by their restraints.

During the design of a space vehicle, egress operations may be tested in idealized environments – test subjects would be healthy individuals suffering no effects of deconditioning and the test spacecraft would generally be unaffected by the dynamics that might be imposed by the atmosphere or, in the case of water landing capsules, the ocean dynamics. Real-world scenarios are not so well modeled. If a crew member is injured during the mission, or is the cause of a medical emergency, that crew member will likely need assistance after landing. Then there also exists the possibility that a crew member (or crew members) may become injured upon landing. And though injuries have happened to flight crews upon landing, there is no way to know a priori what sort of injury might be sustained and by which crew members. Furthermore, there is also no mature way to ascertain the capability of a deconditioned crew member as exposure duration to reduced gravity, susceptibility to prolonged exposure, and individual physiological and psychological response to prolonged deconditioning will be highly variable. Finally, the cramped confines of capsules combined with the potential for a 'hard landing' can (and has) led to the entrapment of a crew member.

This section focuses on how to assess the possibility that a crew member, or crew members, may sustain an injury, entrapment, or other form of incapacitation upon landing and what might be able to be done by other crew members and/or rescuers to respond to the situation safely and efficiently. As injury location and severity, as well as deconditioning, can affect occupant survivability or functionality after mishap, this section also attempts to determine how the egress methods outlined earlier might still be feasible in the case of a post-landing emergency if crew members are injured and/or deconditioned. Finally, the ability to perform the complex operation of an unassisted egress, either suited or unsuited, will be affected by the dynamics of the vehicle and the possibility of the onset of sea sickness, both of which contribute greatly to the challenge of a successful egress in a water landing vehicle.

7.1 Assessing Injury Potential for Egress Operation

If a crew member is injured during the mission, or is the cause of a medical emergency, that crew member will likely need assistance after landing. Then there also exists the possibility that a crew member (or crew members) may become injured upon landing. Injuries complicate or render impossible an unassisted egress operation, which in turn drives overall post-landing survivability. To best assess the potential for injuries upon landing and their effects on egress and post-landing operations, we can review prior incidents of spacecraft landings that have produced injuries to a crew member and also estimate the probability of an injury through analog data sources, such as vehicle crash statistics and auto racing accidents, and through dynamic modeling, as was developed to support the Brinkley Dynamic Response Model.

7.1.1   Prior Incidents of Injured Crewmembers

Though the water landings of US spacecraft have been predictable than the terrestrial landings of the Soyuz spacecraft, many Mercury, Gemini, and Apollo astronauts have described the splashdown event as surprisingly rough. When asked about landing in an interview, Buzz Aldrin of Apollo 11 noted:

> *"It was a lot harder than I expected... I was standing by with my fingers quite close to the circuit breaker. The checklist fell, and the pen or pencil... dropped. It didn't seem as though there was any way of keeping your fingers on the circuit breakers"*

Similarly, Frank Borman, Commander of Apollo 8 mentioned:

> *"The one item that we were perhaps not expecting was the impact at touchdown... There was a severe jolt and we got water in through the cabin repress valves even though they were closed. A good deal of water came in the cabin pressure relief valve"*

Distracted by the sudden surge of water into the capsule, Borman did not activate the pyrotechnic circuits that severed the main parachutes before they pulled the capsule over into a Stable Two.

Apollo 15 landed under only two main parachutes; one had failed to open properly upon descent. Fortunately, the design of the Apollo CM was redundant enough to protect its occupants nominally under such failures and the crew landed unharmed. And though the nitrogen tetroxide contamination of the Apollo CM that was part of the Apollo-Soyuz Test Project was a serious descent and landing contingency, only one injury, that received by Alan Bean of Apollo 12, was reported that was attributable to severe landing impact of an American capsule. The winged design of the Space Shuttle led to much more benign landing characteristics than the capsule designs that had preceded it and no injuries have been reported upon landing by any crew. Soyuz crews, however, have been most susceptible to landing injury mostly as a result of the relative lack of control of the descent trajectory and the unpredictability of terrestrial terrain. Injures have been reported on five Soyuz missions as a result of an anomalous landing – Soyuz 5, Soyuz T-7, Soyuz TM-7, Soyuz TM-12, and Soyuz TMA-11.

Soyuz 5

On January 18, 1969, the parachute lines of the Soyuz 5 DM became partially tangled upon descent, causing a landing 2000km off target in the Ural Mountains near Orenburg, Russia. The force of the landing, exaggerated by a partially closed parachute canopy and the failure of the landing rockets to fire, dislodged the sole cosmonaut from his seat and knocked out several of his top front teeth.

Apollo 12

On November 24, 1969, a minor injury was incurred during the Apollo 12 landing by Alan Bean. The impact was so severe that a camera disconnected from its bracket, striking Bean on his head and rendering him momentarily unconscious. As Bean was closest to the two circuit breakers that would activate the pyrotechnic circuits to free the main parachutes from the spacecraft, it was his job to activate them. But as he was knocked out for about five seconds, the parachutes stayed attached and the capsule flipped inverted, similar to Apollo 8. Pete Conrad, commander of the mission recalls:

> *"We really hit flatter than a pancake, and it was a tremendous impact. Much greater than anything I'd experienced in Gemini. The 16mm camera... whistled off and clanked Al on the head to the tune of six stitches... He was out to lunch for about 5 seconds...I was convinced he was dead over there in the right seat, but he wasn't"* [90]

Soyuz T-7

On December 10, 1982, the Soyuz T-7 DM landed on sloped terrain and rolled down the embankment, eventually settling on its side. In this traumatic landing, the Flight Engineer was ejected from his seat and sustained minor injuries.

Soyuz TM-7

On April 27, 1989, the Soyuz TM-7 DM descended through gusty winds that caused an unusually hard landing. The DM impacted the ground twice and in the process, one of the crew members sustained an injury to his leg that required medical treatment at the landing site before transport.

Soyuz TM-12

On October 10, 1991, the Soyuz TM-12 DM impacted the ground so hard as to significantly dent the vehicle. After impact, the DM came to rest on its side and a cargo canister dislodged, striking one crewmember on his head and causing an injury that was treated at the landing site.

Soyuz TMA-11

On April 19, 2008, the crew of Soyuz TMA-11 experienced a ballistic re-entry of over 8 G's force following a delayed separation of the DM from the SM, which produced smoke in the cabin, a failure of the soft landing system, and a very hard landing. They landed 470 km short of the target point, igniting a small grass fire at the landing point. The crew was injured and had to be helped from the DM by nearby residents. Cosmonaut Yuri Malenchenko and Astronaut Peggy Whitson suffered no permanent injury, but Korean Cosmonaut So-YeonYi was hit by Whitson's personal effects bag on impact and required physical therapy for neck and spine injuries.

7.1.2   Assessing the Probability of Injuring or Incapacitating Crewmembers

The possibility of an injury upon landing is much greater in a capsule spacecraft than in a winged spacecraft. There has never been an injury sustained by landing throughout the Space Shuttle program, and it may be safe to say that if a winged spacecraft was to follow a nominal approach, and in the absence of a failure of the landing gear system or any other critical mechanical component, the probability of injury that a crew might face due to impact upon landing would be comparable to that of a crew of an aircraft landing on a suitable runway. If, however, the same winged spacecraft failed to maintain a proper approach and were to land off the runway, the higher approach and touchdown speeds that winged spacecraft currently have compared to aircraft drive significantly the possibility of injury or a LOC event.

The landings of capsule-based spacecraft are less predictable. The impact velocity becomes a function of the effectiveness of redundant systems designed to reduce the landing velocity, the angle of impact relative to the terrain and, arguably, the physiological state of the crew. Still, parachute descents place the crew at the mercy of winds and other environmental factors and even the most stark terrains that are often selected as landing zones still contain hazards.

Variables

Numerous variables exist in the assessments of injury risk through impacts as might be received through capsule-based spacecraft landing. The anthropogenic measures and gender of the astronauts should be considered. The Space Shuttle is currently designed to accommodate the 5th percentile female to the 95th percentile male. The anthropogenic bounds of the Soyuz spacecraft are more constrained. Injury risk is also affected through the physiological effects of long duration spaceflight deconditioning, which may lead to increased risk for certain types of injuries. In the interface between the occupant and the space suit, injury risk may be elevated through point loading in the suit design, blunt trauma from suit interaction, and interactions with restraint mechanisms integral to the suit. In the interface between the suit and the seat, protection against flail, or movement caused by the impact, must be adequate.

Analog Data Sources

Assuming then that the risk of the presence of an incapacitated crewmember during recovery is a function only of the reliability of the landing systems, we can best model this through analogy with the Soyuz program. As of December 2011, there have been 123 Soyuz spacecraft landings, excluding the pad abort of Soyuz 10-1 and the ascent abort of Soyuz 18-1, of which two (Soyuz 1 and Soyuz 11) involved a LOC event and five produced injuries (as detailed prior). From these statistics, a rough order-of-magnitude estimate of injury probability for a capsule-based spacecraft landing on land would be about 4%. The major drivers of injuries were 1) higher than nominal impact velocities driven by partial deployment of the parachute system or failure of the landing rockets and 2) gusty winds at the landing site.

But spacecraft landing statistics are still rather sparse statistically. Despite the dearth of data from space missions, there nevertheless exist large quantities of relevant analog data that can provide value to any analysis of landing injury risk. The unfortunate reality of our mobile society is that air and land vehicles do have accidents that produce injuries and fatalities with regularity. The National Transportation Safety Board (NTSB) maintains an aviation accident database contains information from 1962 and later about civil aviation accidents and selected incidents within the United States, its territories and possessions, and in international waters. In addition, the Crash Injury Research Engineering Network (CIREN) and the National Automotive Sampling System (NASS) maintain databases involving accidents of passenger vehicles. The mission of the CIREN is to improve the prevention, treatment, and rehabilitation of motor vehicle crash injuries to reduce deaths, disabilities, and human and economic costs by determining the injury causation in every crash investigation conducted. NASS maintains data on passenger vehicle crashes in order to investigate injury mechanisms that can be used to identify potential improvements in vehicle design.

Probably the best proxy that can be used as an analog to spacecraft landing impacts comes from the sport of auto racing. Accidents in auto races are fairly common, the drivers well protected through the best engineering methods available, and data is meticulously recorded. In the US, the Indy Racing League (IRL) and the National Association for Stock Car Auto Racing (NASCAR) have constructed the largest datasets in the interest of improving driver safety.

EGRESS AND POST-LANDING OPERATIONS IN THE AGE OF CREWED COMMERCIAL SPACE FLIGHT

NASCAR has compiled statistics on 4071 actual impacts incurred during races and performed an injury biodynamics analysis using dummy responses along with actual injury data to develop a probability of injury. Figure 7.1 illustrates through a polar representation the distribution of injuries received by a driver in a NASCAR race as a function of the magnitude of the impact and the angle from which the driver's car was struck. We see in this representation that injuries have been produced from even the most minor impacts, yet there is greater probability that a driver will sustain an injury if struck from the side.

In addition to data sets provided through automobile racing, another source of analog data is provided by through the operational history of ejection seats. Unlike the rapid deceleration of a vehicle crash, ejection seats produce a rapid, and often violent, acceleration to avoid a potentially catastrophic scenario.

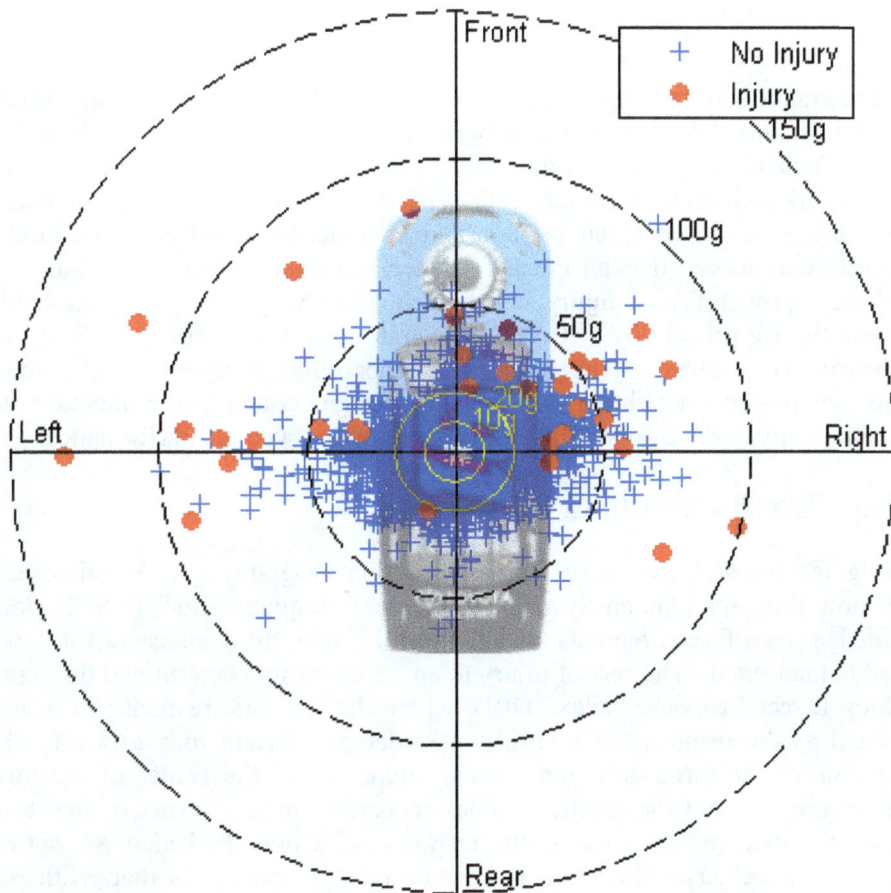

Figure 7.1 NASCAR Injury Trends (source: NASCAR)

<u>Developing Models</u>

The NASCAR data, along with data from other sources, are used by NASA to develop analytical tools and methodology that can be used to define better injury criteria and human tolerance limits that can better assess injury potential and develop more accurate human tolerance limits applicable to spaceflight. To do this, human impact testing is conducted with crash dummies in mock-up spacecraft in relevant environments which can use existing crash test dummy models in vehicle specific restraints. Real-world data from NASCAR and other sources is used to "normalize" the dummy responses in crash scenarios to injury. Crash dummies (e.g. the Toyota THUMS human model) can also be used to investigate the potential for suit-related injury, such as through blunt trauma, point loading, or restraint interactions.

Human Injury-Risk Modeling

Impact deceleration on landing is set so that the risks of incapacitating injury are minimal. The standard method to quantify injury potential due to impact is defined by the Brinkley Dynamic Response Model, which can be adjusted to account for an injured and/or deconditioned crew member. The Brinkley Model treats the human as a mechanical system exposed to an impact, and attempts to establish a threshold value where injuries can occur. It is an inexact science and open to a degree of subjectivity. Nevertheless, a probability of injury is established for very low (injury probability < 0.05%), low (injury probability < 0.5%), moderate (injury probability < 5%), or high (injury probability < 50%). The lower the probability of injury sought, the more constraints are placed on vehicle and mission design, which could increase weight, increase complexity, reduce operability, and reduce overall vehicle performance.

<u>The Brinkley Direct Response Index</u>

While designing the architecture of the Constellation Program, NASA had released the "Constellation Program Human-Systems Integration Requirements" (HSIR) document that detailed a set of requirements which must be met by a spacecraft for it to be considered human-rated. The risk of injury to an occupant was determined through use of the Brinkley Direct Response Index (DRI), where the dynamic response of a human is approximated as the response of a spring-mass-damper system to a given acceleration profile in each of the three orthogonal axes, referenced to the center of the torso and shown in Figure 7.2, and the limiting values represent varying levels of risk to injury. Simply put, the dynamic response is the response of a hypothetical mass, spring, and damper system attached to the seat. The virtue of this concept is that, with properly selected coefficients, the peak acceleration of the mass is monotonically proportional to the probability of injury. A different dynamic model is used for each orthogonal axis

The HSIR assumes that only linear and rotational accelerations exceeding the threshold value incurred upon landing may contribute to the probability of the presence of an incapacitated crewmember during the recovery operation. This Dynamic Response Model established for Orion as it had for Apollo a low-probability of injury upon parachute deployment or landing (a risk of sustaining a serious or incapacitating injury of no greater than 0.5%) for a healthy deconditioned and/or an Injured crewmember. Multiple off-nominal failures could impart risks in the medium risk and high risk categories (5% and 50% risk of sustaining a serious or incapacitating injury). Examples of off-nominal conditions are (i) a landing with one parachute failed, and (ii) a landing with a component failure in the landing attenuation system.

Though the Constellation Program initially insisted that all vehicles must maintain the level of injury-risk to its occupant in the "very low" range throughout a transient acceleration event for it to be considered human-rated, the requirement was later relaxed to a "low" limiting injury risk level. Higher levels of acceptable risk are justified for contingencies involving failures of key systems.

To help determine the percentage that such 'hard-landings' would cause a crewmember to be incapacitated, Figure 7.4 through Figure 7.7 illustrate the crew load limits for nominal and contingency impact events, imposed on a coordinate reference system as depicted in Figure 7.2. The solid red lines in these figures represent the maximum level of sustained acceleration allowed on a crewmember during a launch abort or emergency entry. Under these extreme conditions, it may be necessary to expose the crew to accelerations more severe than those experienced nominally (see dashed blue lines), but crewmembers should never be exposed to accelerations greater than those depicted by the solid red lines in the charts. Exceeding these elevated limits could significantly increase the risk of crew incapacitation, thereby threatening crew survival.

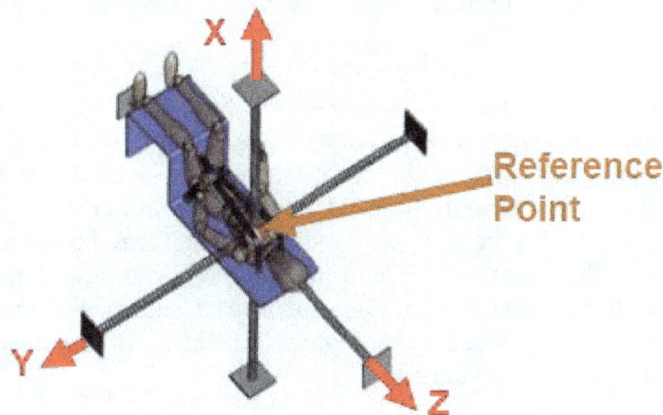

Figure 7.2 Brinkley Reference Coordinate System

Impact velocities and angles

Spacecraft landing systems are critical systems that can quickly lead to a LOC event and are thus redundant so that the crew can survive the failure of a major component; the

Apollo CM was designed to protect the crew if one of the three parachutes were to fail and the Soyuz DM contained a reserve parachute which was designed to protect the crew even under a failure of the landing rocket system.

Soyuz TMA Reentry Configuration	Speed of Descent	Delta-t	G-Force
Main Parachute Deployed	7.41 m/s	0.135 s	110 g's
Main Parachute Failure, Reserve Chute Deployed	8.24 m/s	0.121 s	137g's
Main Parachute Deployed, Landing Rockets Fired	1.57 m/s	0.637 s	5.9 g's
Reserve Parachute Deployed with Landing Rockets	2.40 m/s	0.417 s	12.5 g's

Table 7.1 Soyuz Reentry Speeds under Various Configurations

As an example, the various landing velocities with the associated time of deceleration (delta t) and corresponding accelerative deceleration (g-force) are tabulated in Table 7.1 for the Soyuz TMA DM. The speed of descent, combined with the effectiveness of the seat and strut system to absorb an impact, determine the delta-t and the G-force the crew member will experience. Though the reserve parachute is not as effective as the main parachute, both slow the DM sufficiently if the landing rockets function properly. If the landing rockets were to fail, the resultant impact would be survivable, though the impact would be in excess of HSIR limits as defined by NASA, and it is likely the crew could be injured.

Just as sloping terrain can cause a capsule-based spacecraft to impact the land at an unanticipated angle, the dynamics of waves in the ocean may cause the vehicle to impact at an angle anywhere from zero (at the crest or trough of a wave) to a maximum value midway between a crest and trough. Impacts on sloped terrain or water surfaces distribute the forces (as shown in Figure 7.3) in ways different from how the seats were best designed to protect the occupant. A well designed seat and landing system will sufficiently protect the occupants at a range of impact angles that the spacecraft could reasonable expect to be exposed to. Figure 7.3 shows a capsule-based spacecraft impacting the landing site at varying angles.

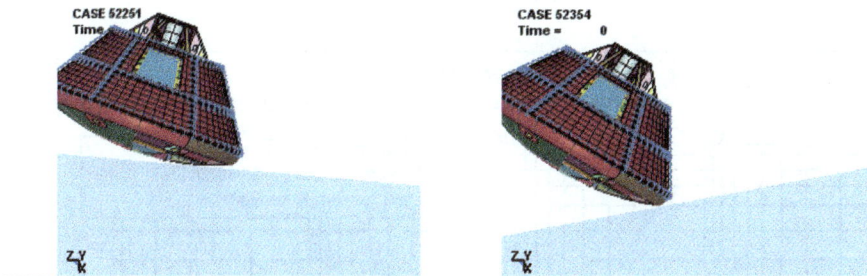

Figure 7.3 Orion spacecraft landing under varying angles of impact

Rescue Operations involving Injured Crewmembers

As a rule, and barring the presence of a post-landing contingency, recovery methods which keep the crew contained within the spacecraft pose less risk than recovery methods requiring crew extraction.   But in the preparation of recovery and egress scenarios involving injured crew members, two cases are generally considered: 1) one crewmember is injured and can expect support from other flight crew members, and 2) all crewmembers are non-responsive when the recovery team arrives. For situations where one crew member is considered incapacitated, the other crewmembers may provide assistance. If all crewmembers are incapacitated, the rescue effort would need to be conducted solely by external rescuers.

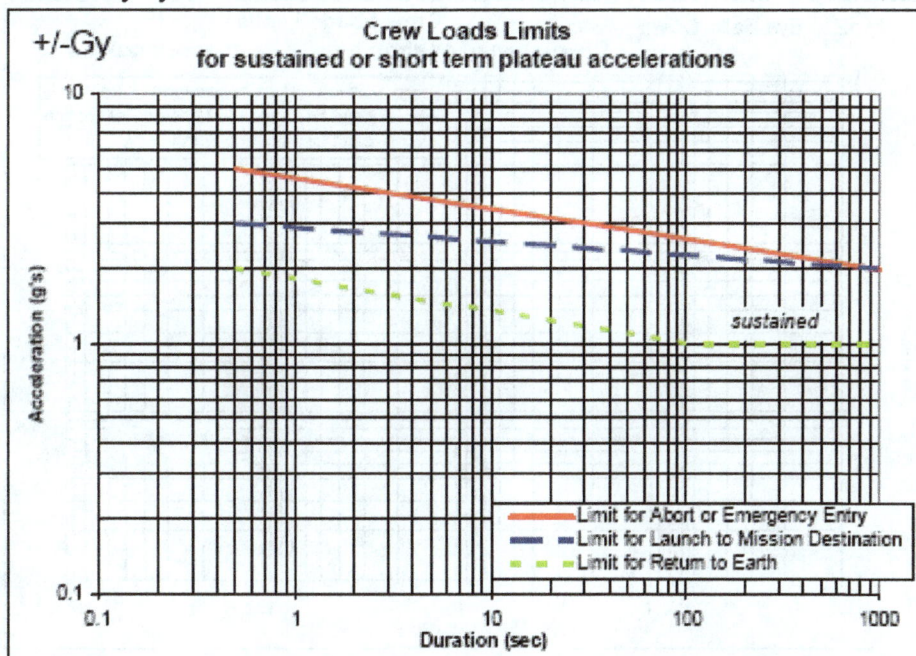

Figure 7.4 HSIR +/- Gy Crew Load Limits

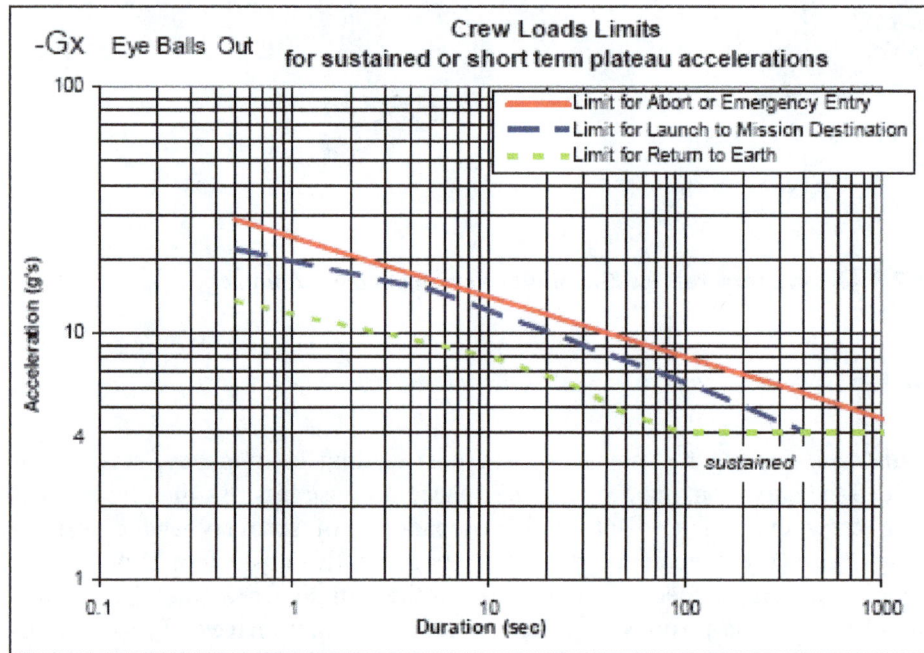

Figure 7.5 HSIR -Gx Crew Load Limits

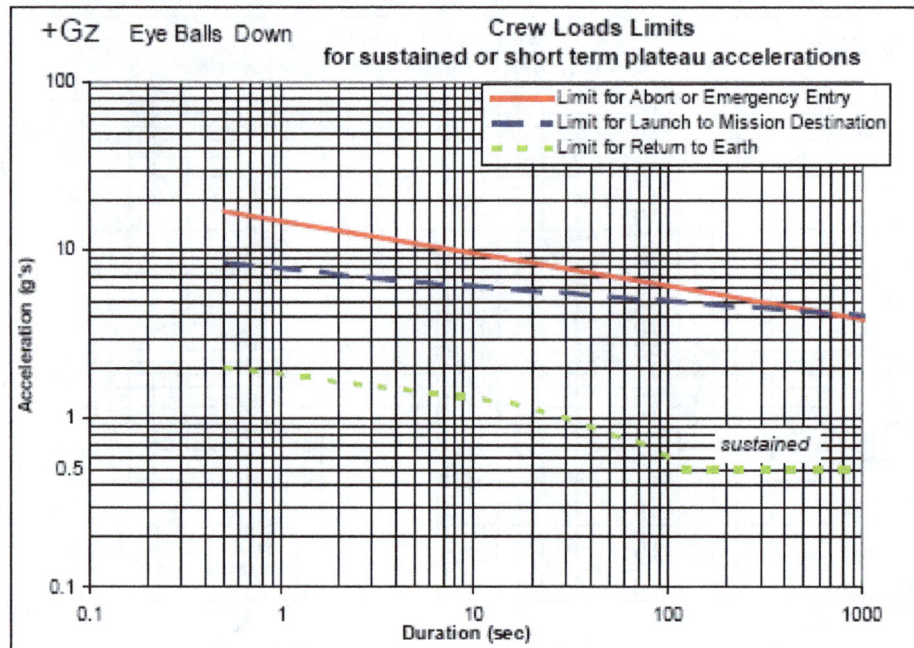

Figure 7.6 HSIR +Gz Crew Load Limits

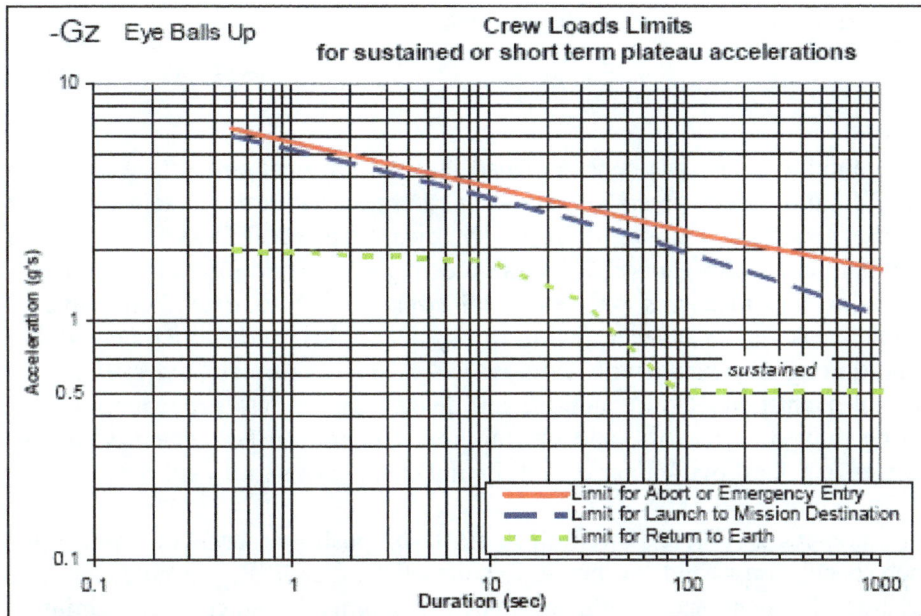

Figure 7.7 HSIR -Gz Crew Load Limits

7.2 Assessing the Effects of Deconditioning on Egress Operations

NASA astronauts typically wear about 91 pounds of equipment upon landing. A crewmember not properly trained may be ineffective during egress and require excessive time in a dangerous environment. Further, this additional encumbrance could make it extremely difficult or impossible to perform an emergency egress when in a physically deconditioned state that would follow long exposure to reduced gravity. According to a NASA flight doctor, no reliable results currently exist that would determine how each pound of mass, i.e., crew escape equipment, might impact the mobility of a deconditioned crew member upon re-introduction to a gravity environment.

The extent of deconditioning caused by microgravity will vary based on the size and strength of each crew member and the amount of time spent on-orbit. Data from deconditioned astronauts is rare, and much concerning operational feasibility is gleamed from 'expert opinions' from those that have been in a state of deconditioning. Aside from these opinions, the best source of analog data comes from participants in 'bed rest' studies. Bed rest studies are "the most widely used simulation of the space flight environment, causing decreased orthostatic tolerance similar to that demonstrated by returning astronauts", according to "Space Physiology and Medicine". Any post-landing procedures, especially emergency egress procedures, should be validated both through the

179

use of bed rest subjects in full-scale egress tests as well as by the assessment of astronauts that have performed long-duration spaceflights.

7.2.1 Physiological Effects of Deconditioning

Limited quantitative data has been obtained through physical strength tests administered upon long-duration space flyers shortly after returning to Earth, largely but not entirely through the Extended Duration Orbiter Medical Project (EDOMP). The EDOMP collected data on 18 astronauts that had all recently completed long duration missions on the ISS. The test measures only physical strength and to date there have been no quantifiable measures for other measures of physical fitness and endurance and the effects of deconditioning associated with cardiovascular, neurovascular, and skeletal condition are not well quantified. Consequently, suit designs and egress procedures are designed considering only strength loss, and the failure to be able to perform procedures due to other factors (e.g. cardiovascular deconditioning) remains a significant risk.

The test compares a subject's strength in six tests: push-ups, crunches, pull ups, the sit and reach, bench press, and leg press. Results of the EDOMP, presented in Figure 1.1, indicate that the post spaceflight estimated strength decrement for minimum crew operational loads vary between zero and 26%, with an average decrement of 20%. Post-flight data and ground (bed rest) analog testing data all suggest that 20% is a valid estimation for muscle strength reduction.

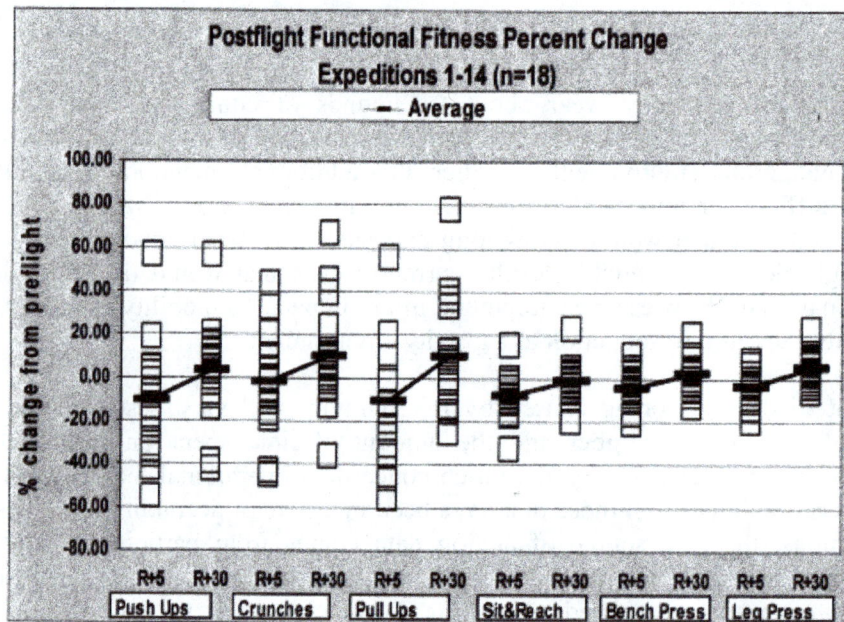

Figure 7.8 Post-Flight Functional Fitness Percent Change due to Deconditioning

Specifically, musculoskeletal strength was observed on average to decrease 20%. The lower extremity strength showed a much higher relative strength decrease of 40%. Paravertebral and spinal effectiveness also decreased 40% on average.

In addition to temporary reductions of muscle mass and physical strength, long duration spaceflight has been observed to cause longer lasting effects. Bone mass decreased as well, with weight bearing bones decreasing in mass approximately 1-2% for each month on orbit. Long duration spaceflight affects cardiovascular performance as well. Cardiac muscle size decreased as well, as being on-orbit for 6-months has been observed to create an intravascular volume loss between 6 and 12%. Recent studies have also indicated that long-duration spaceflight can cause a spectrum of changes in astronauts' visual systems, including blurry vision, that are believed to be caused increased intracranial pressure.

Upon landing, deconditioned crewmembers have reported sensations of nausea and vomiting associated with the onset of a gravity environment while being in a state of deconditioning. Rapid head movements can cause a risk of incapacitation. These effects may not have long-lasting implications but may complicate an astronaut's ability to perform an egress operation in an emergency. Deconditioning may render a generally healthy crewmember unable to egress their seat for several hours and may make it impossible to render aid to another crewmember or perform an emergency egress.

Water landings are particularly demanding on the crew post-landing, and deconditioning plays a major contributor to risks associated with emergency egress. To date, the longest duration that astronauts have been exposed to reduced gravity prior to a water landing has been the crew of Skylab 4, consisting of Alan Bean, Jack Lousma, and Owen Garriott, who remained on orbit a total of 84 days. Durations of the three Skylab missions, all involving crew return to a water landing in an Apollo CM, are listed below:

Skylab 2 (SL-2)	May 25, 1973 to June 26, 1973	(28 days)
Skylab 3 (SL-3)	July 28 1973 to September 25 1973	(59 days)
Skylab 4 (SL-4)	November 16, 1973 to February 8 1974	(84 days)

Since Skylab, many astronauts and cosmonauts have logged longer times on orbit, but all have returned to a terrestrial landing in either the Space Shuttle or in the DM of a Soyuz spacecraft. Crewmembers spending six months or more on the ISS and returning on a water-landing vehicle, as the Orion capsule was to do, may find themselves unable to effectively respond to a post-landing emergency.

The Russian space program estimate total body musculo-skeletal strength decrements of 20% to 40% after long duration space missions as well as bone mass reductions of 3% to 18% for weight bearing bones and risks of spinal column injury (e.g. herniated discs). They recommend that all crewmembers should be treated as "unconscious un-witnessed trauma patients" and given spinal stabilization with full body restraints. Further, crewmembers are instructed to avoid any head movements, as these would cause dizziness, nausea and vomiting, and possibly incapacitation due to inadequate neuro-vestibular re-adaptation to a gravity environment. Cardiovascuar system will be affected as well by an intra-vascular volume loss of 6 to 12% coupled with decreased cardiac muscle and filling capabilities. Signs of decompression sickness may also result from a contingency during re-entry. As a precaution, pure oxygen is administered and any air transport of crewmembers is restricted to low altitude (<1000 feet AGL).

7.2.2 Psychological Stressors Associated with Deconditioning

In addition to the physiological effects of reduced gravity deconditioning, long term spaceflight introduces numerous stressors that affect the psychological state of crew members. A post-landing emergency may require a clear and decisive decision making capability to avert a hazardous situation. Aside from the psychological stresses that will inevitably accompany the physiological stresses of a deconditioned state, crews returning from long-duration spaceflight also are affected by long periods of confinement, isolation, and interpersonal and environmental challenges. Interpersonal challenges may arise from cross-cultural issues and interpersonal tensions as well as sensory depravation and lack of privacy. These psychological stressors can grow to create physiological stressors such as sleep disturbances, headaches, irritability, anxiety, depression, restlessness, and loneliness. In turn, these stressors may diminish a crew member's capability to effectively make and execute time-critical decisions in a time of emergency.

7.2.3 Additive Effects of Sea Motion and Sea Sickness

Motion and the effects of sea sickness can play a significant part to egress effectiveness of a crew from a water-landing vehicle. Obviously, rougher seas will make egress more difficult and dangerous. Further, if a crew member is succumbing to the effects of sea sickness, this will also make them more ineffective.

As these analyses involve many human factors issues, they are intrinsically difficult to quantify, but scaled and analytical models can adequately determine the dynamic behavior of a vehicle in water. The US Navy has also constructed motion tables and sea sickness indices to estimate the effectiveness of individuals performing operational tasks while under the influence of an oscillating, dynamic environment and/or the effects of sea sickness.

In tests, good results can be obtained in a wave pool. Just as the use of test subjects deconditioned through use of bed rest can make a good proxy to space deconditioned crew members, wave pools and other dynamic motion generators may provide a cost effective means to simulate motion. The best analog test would involve well bed-rested test subjects that have been in strapped into a boilerplate test vehicle that is floating in a wave pool. The test subjects may then experience additive effects of motion sickness. Nighttime lighting conditions may also be tested, though it may be assumed that adequate lighting exists within the vehicle.

Though this test would be the best analog, in reality it is expensive to perform. In the absence of bed-rested test subjects, the next best way to create meaningful results though a test is to perform the test in an idealized environment and determine a realistic multiplier to account for the more subjective effects of vehicle motion, sea sickness, and deconditioning. The egress times expected for emergency egress from the Orion capsule were determined in this manner where a consensus of formerly deconditioned astronauts and space medical officers agreed upon a multiplicative factor of 3.0.

7.2.4   Pre-Landing Countermeasures

To best prepare a returning crew for the stresses of landing and the possibility of performing a strenuous post-landing operation, several countermeasures may be taken. To get the crew in optimal physical condition, a period of fluid loading precedes the de-orbit burn. The crew would then doff suits that have anti-G properties, though these suits may limit rapid medical access. Upon landing, each member of the crew would have received extensive egress training and survival training to avoid unnecessary accidents and also to make every effort in an egress operation as intentional and effective as possible. The first response of the rescue team should be as rapid and capable as possible and a medical operations contingency plan should be intact.

7.3 Incapacitation through Entrapment

There has also been historical precedence of a crew member becoming trapped. The Soyuz TM-14 DM suffered a hard impact upon landing on August 10, 1992 as a result of high winds. As a result, the DM turned over on its side and the crew was left lying sideways and the escape hatch was jammed so that outside rescue teams could not open it, though the crew was eventually successful in opening the hatch from the inside. As a result of the hard impact, one crewmember was trapped in his seat as his helmet became jammed with cables that had dislodged from their harnesses. The crewmember was rendered immobile until his suit was eventually cut free from the rescue team.

Though it is difficult to predict how a crew member might become entrapped within a spacecraft, this example demonstrates the importance of thorough impact and shock testing for all impacts that might be caused in both nominal and off-nominal landings. Human factors concerns, including the effectiveness of an egress operation, should be considered as these tests are performed.

7.4 Available Medical Resources

Medical equipment is relevant to post-landing operations, especially if the recovery is delayed. Crews in a water-landing capsule are particularly susceptible to sea sickness. Early spacecraft had to carry medical kits through the entire mission. With most space missions currently supporting the ISS, medical supplies may be much more limited on the vehicles used for transport to and from the station.

The first medical kits supplied to astronauts on Mercury capsules were sparse kits that contained only three medications to serve as stimulants and to mitigate motion sickness. Subsequent, longer missions were better prepared. Since no preventive medicine program, however carefully conceived, can ever guarantee the absence of illness or disease, medications were carried onboard the Gemini and Apollo spacecraft, though the contents of the medical kit were constantly revised as need indicated. The Shuttle Orbiter Medical System (SOMS) can treat minor illnesses and injuries as well as provide support for stabilizing severely injured or ill crew members until they are returned to Earth. The ISS currently provides 191 medications that can be administered in numerous ways and has a similar capability to treat medical emergencies as one might find on a remote polar outpost. With all space missions, diagnosis and treatment of injuries and illnesses is performed through consultation with flight surgeons on the ground.

The subject of space medicine is far too involved to broach within the scope of this book, but the following lists of medical supplies for Gemini missions and Apollo missions may provide a good insight into what sort of oral and injectible medications may be available within a typical spacecraft upon and after landing. In addition to medications, Apollo crews had available to them a clinical monitoring system which included a stethoscope, thermometers, and a blood pressure cuff. A biomedical instrumentation kit also provided sensors, micropore disks, wet wipe cloths, and electrode paste.

Gemini VII Medical Kit Contents

- Oral Marezine   (8 x 50mg pills for motion sickness)
- Parenteral Marezine   (2 x 45mg pills for motion sickness)
- d-Amphetamine sulfate   (8 x 5mg pills, as a stimulant)
- Aspirin, Phenacetin, Caffeine   (16 pills)
- Oral Meperidine HCl   (16 x 100mg pills for decongestant)
- Parenteral Meperidine   (2 x 90mg pills for pain)
- Actifed 2.5mg   (16 pills as a decongestant)
- Sudafed   (16 x 60mg pills as a decongestant)
- Lomitil   (16 x 2.5mg pills for diarrhea)
- Tetracycline HCl   (16 x 250mg pills as an antibiotic)
- Methyl cellulose solution   (1 x 15mL eye drops)

Apollo Medical Kit Contents:

- Bismuth sub-carbonate   (24 pills, fever, pain reducer)
- Darvon compound 65   (12 pills, fever, pain reducer)
- Globaline   (50 pills, suppress infection of GI system)
- Tigan, Bomodoxin, or Marezine (24 pills and 3 injections, anti-nausea)

- Dexedrine   (12 pills, stimulant)
- Acromycin   (24 pills, antibiotic)
- Morphine Sulphate   (3 injections, pain killer)

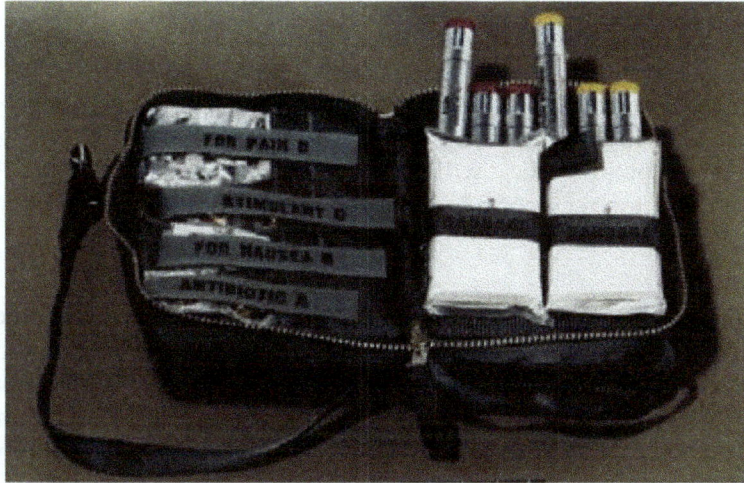

Figure 7.9 Apollo medical kit (source: Smithsonian Institution)

8 Egress and Post-Landing Operations in the Age of Commercial Crewed Spaceflight

The future brings more than vehicles of national space programs. Soon, international crews will orbit the world comprising crew members, scientists, artists, educators, or simply the interested public. How will contingency rescue plans be integrated into the mission planning of these new commercial spacecraft? The answers will be political as well as technical.

But how will the next generation of spacecraft land? What will the most credible failures be and into what environments will the mitigation methods of these failures bring the crew? And what can happen after landing and how can the crew best prepare for these events? These questions are being considered certainly with lots of issues 'to be decided' as they are questions involving lots of human factors issues and randomly distributed failure and environmental data. It will remain, for a long time, an inexact science. But as spacecraft missions multiply, the infrastructure will expand. Spacecraft will have available more suitable landing facilities and more mature contingency response plans available. Landing systems will mature; retro-rockets will become more reliable; the envelope of stresses, vibrations, acoustical buffeting, and g-loads the human occupants will be exposed to will shrink; ascent aborts will become rarer and the ramifications of aborting will no longer be a life-threatening 20-g event but rather something much more manageable.

For now and the foreseeable future, ascent aborts will remain a major concern. Capsule based designs have an intrinsic randomness to them. The Soyuz spacecraft shows no signs of abating in its role as a reliable transporter of personnel to low earth orbit and the *China* National Space Administration is building their own vision of a national space program about the heritage of the Soyuz spacecraft. The Orion capsule, designed initially as the crew module of the Constellation Program that was cancelled in 2010, is being reborn through commercial enterprises using its heritage design such as the SpaceX Dragon and the ATK Liberty vehicle.

But new national programs are being born as well. Russia sought participation with the US during formulation of the Constellation Program, but was denied. As a consequence, the European Space Agency (ESA) teamed with the Russians to develop a joint project. The ESA is in process of developing the Crew Space Transportation System (CSTS), which will initially service the ISS but also be able to conduct exploratory missions of the Moon and beyond. Meanwhile, in Russia, the development of the Russian Prospective Piloted Transport System (PPTS) will aim to replace the aging Soyuz system. Both of these designs share similar mission objectives, and at this time both borrow heavily upon the design characteristics of the Orion spacecraft.

There are new entrants into the realm of national space programs as well. India is developing their own crewed space vehicle that is of a capsule design, and the low latitude east coast of India lends itself perfectly to the establishment of a domestic spaceport. Japan and Iran are also seeking their own crewed space programs, presumably using capsule designs.

The United States will present the initial arena to determine if the economic climate can spur, with the right technology infusion and public investment, a private crewed space industry. In a July 10 interview on The Space Show, Dr. George Nield of the Federal Aviation Administration's Office of Commercial Space Transportation (FAA/AST) appeared to describe their position as essentially *laissez faire*; that it is largely up to the suborbital operators to determine their policies and procedures.

It is interesting to note the role of the FAA during the emergence of commercial spacecraft and the parallels of the experimental aircraft industry. In the early 1980's manufacturers of general aviation aircraft saw their industry virtually collapse under the burden of extensive regulations, fuel crises, and especially the sharp increase of product liability lawsuits. The rebirth of the industry came in through the hackers and homebuilders that built a new breed of flying machines in the confines of garages and hangars. The FAA established a principle of informed consent of those that would ride in such vehicles and a principle of protecting the uninvolved public. They FAA allowed the construction, testing, and operations of these vehicles provided that people can assume the risks and understand the dangers involved. As a result, a plethora of new designs emerged – many relying on new technologies in composite design, avionics, and propulsion systems. Eventually, the best designs rose to the top and viable, profitable companies formed. Of these, the best vehicles underwent the process of certification and emerged in the late 1990's as the new generation of general aviation aircraft.

The Commercial Space Launch Amendments Act of 2004 (CSLAA) requires commercial spaceflight to be essentially as safe as commercial aircraft operations. But accidents will happen - it is only a matter of when they will happen. If an accident occurs with damage to uninvolved people or property, or injury to commercial passengers, the entire industry could be jeopardized, and the commercial spacecraft developers surely recognize this and share a mutual responsibility to create an industry-wide climate of safety. A means to an effective egress will be integrated into each design, but how? And what will the spacecraft operators be required to provide in terms of crew and passenger training?

8.1 Commercial Design Considerations for Egress

The first crewed commercial space vehicles will likely be suborbital winged spacecraft, such as those being produced for Virgin Galactic and those being produced by XCOR. These vehicles that follow a ballistic trajectory after MECO and glide back to a landing at the same place from which it was launched. If a problem were to be detected that would cause an engine cut-off, the crew might initiate a runway abort or glide to an early landing. After MECO, the vehicle is essentially a glider and if any failure were to happen in this stage, the crew would most likely remain in the vehicle through landing, unless the vehicle was unable to land safely (e.g. through failure of the landing system, inability to return to the runway, etc.). In this case, a bailout egress could be required once the vehicle could decelerate well below Mach One. Even if cabin pressure is lost, remaining in the vehicle would still be preferable since the cabin provides some protection against the high stagnation temperatures encountered during the return to denser atmosphere.

As the suborbital market expands, there will be a demand for point-to-point suborbital flights. Logistical planning for such flights would involve the identification of suitable runways at each point of the runway. If there were to be any times at which an engine cut-off would place the spacecraft outside of safe gliding distance to a suitable runway, then the most preferable landing area would need to be identified and this might force the crew to decide between performing a controlled landing at a sub-nominal landing area or bailing out of the vehicle.

Initial crewed commercial vehicles to enter orbital flight will likely either be air-launched, such as the system proposed by Stratolaunch LLC, or they will be vertically launched, such as the SpaceX Dragon or the Sierra Nevada Dream Chaser spacecraft. For a vertically launched capsule-design will need an abort system, and an abort system that can terminate engine thrust and get the crew away while still in the spacecraft via a LAS system has definite advantages over ejection seats that might expose the crew to a fireball. But larger winged spacecraft might not be able to eject the crew in this manner, in which case ejection seats may be the best means to egress. Ejection seats, however, have a much narrower window of operability in terms of altitude and velocity. For horizontally launched vehicles, ejection seats may be a viable option to evacuate the crew from an emergency that does not involve a catastrophic failure of the vehicle (e.g. explosion). As the vehicle accelerates and pitches upwards, ejection seats lose their effectiveness. In general, ejection seats are of limited value above Mach 0.9 at sea level to perhaps Mach 3.7 at 20,000 meters because of high dynamic pressures and/or stagnation temperatures. Above 20,000 meters, high stagnation temperatures are problematic for survival without a capsule above speeds of about Mach 2.5 to 3.7. It will be these spacecraft that will need to borrow the most in terms lessons learned involving landing, egress, and post-landing operations from the preceding vehicles of national space programs, but each vehicle and mission profile will have its own unique distribution of risk, its own distribution of landing scenarios, and its own cost and weight versus risk-mitigation trades to be conducted.

Eventually, technological advances will allow spacecraft to better target specific landing areas. Companies such as SpaceX and Blue Origin are currently developing vehicle and booster capability to land vehicles and boosters following a controlled descent to a landing at a specified location. Such systems would be costly in terms of the added fuel mass that would burden the vehicle but would have large benefits in mitigating overall risk and reducing mission costs through greater reusability of vehicle components.

Further, if launch vehicle boosters can be safely recovered through an active powered landing system, the FAA might be convinced one day that such boosters would not pose a threat to the general public and loosen the regulations involving populated areas near the ascent groundtrack. If this is so, then crewed space launches may not need extend over water and instead could launch from large, unpopulated regions where specific terrestrial landing areas could be identified along the ascent groundtrack that could support the landing of a powered capsule that could translate itself away from hazardous areas. If this technology matured to a high state of reliability, then much of the risk associated with water landings or untargeted terrestrial landings would disappear.

8.2 Training to Egress from Commercial Spacecraft

As egress methods and systems are only as good as the efficiency of the procedures the crew would perform, and the procedures will be performed as efficiently as the crew has been trained to do, the issue of egress training remains one of critical importance to overall mission safety. But for crewed commercial space vehicles, what would the requirements be and who would establish and maintain them? The FAA has yet to establish any requirements concerning emergency egress from commercial space vehicles, so the spacecraft manufacturers are left with the task of defining the standards to be complied with for their own vehicles. Until the designs of each commercial spacecraft is finalized, spacecraft egress training will be managed in a way that builds only generalized skills – skills that could be applied to any air vehicle that might suffer an off-nominal landing. Once a final design is established, boilerplate spacecraft could be constructed to best train crews to conduct and direct an egress operation.

Suit design, of course, plays a key part on overall egress planning, and at least one commercial space operator is considering the use of pressure suits for its passengers. Just as passengers on commercial aircraft no longer wear parachutes, there will come a day when the risks of a cabin depressurization would be low enough to accept the risk of having the passengers unsuited. But in the initial stages of commercial crewed spaceflight, we can assume that pressure suits will be commonly used to reduce the risk associated with sudden cabin depressurization. If a cabin decompression were to happen above 10,000m, the crew would have only seconds to identify and fix the problem, lest they suffer the same fate as the crew of Soyuz 11. Pressure suits mitigate this risk, but introduce mobility limitations on their wearers. Pressure suit designers thus prefer to keep the pressure differential between the inside and the outside of the suit as small as possible (the shuttle Advanced Crew Escape Suit, or ACES suit, was pressurized to about 222 mmHg). In addition to protecting against the risks of a sudden depressurization, suits also protect against decompression sickness. A survey of more than 400 U-2 pilots, whose cabin environment was designed to maintain an equivalent altitude of 8,800–9,100 meters (29,000–30,000 feet), found that many reported symptoms of decompression sickness during their careers, and more than 10 percent reported that they altered or aborted their missions as a result, though most symptoms disappeared upon return and after breathing pure oxygen for a limited period. Supplemental oxygen systems that interface effectively with the spacesuits would be an essential backup system in case of cabin depressurization, though they add one more element to encumber a crew member in the event of an emergency egress operation. Because of this, adequate training should be provided to flight members on suit use, with and without supplemental oxygen, including egress operations while wearing the suit, prior to an actual mission.

Finally, just as properly trained air crews know the basics of survival within the terrain that they routinely fly over, commercial spacecraft crews should have an adequate knowledge of survival within the environments that they will fly over. For orbital missions, these environments would be those environments confined by the limits of the orbital inclination. For example, a spacecraft orbiting along a 30-degree inclined orbit would have no possibility of landing in a polar environment, but should have adequate sea and desert survival skills. Wilderness survival training available to aircrews is readily available and would translate easily to spacecraft crews.

Just as post-landing training is connected with the response time and capability of the rescue assets, and available assets depend on where the spacecraft might land, training of flight crews must be general enough so that the crew could be as adaptable as possible to any potential environments and be prepared to interface with a wide variety of recovery assets. On the other hand, SAR and first response teams that may find themselves called upon to assist a spacecraft in distress should be able to rapidly inform themselves of the specifics of the rescue operation they will engage in. Spacecraft can provide unique hazards to the rescuers as well as the flight crew; toxic propellants may still outgas from thrusters and hatches may be pyrotechnically activated by a crew unaware that personnel might be in harm's way. If the vehicle were to land in water, it is a well-known fact that spacecraft make poor boats and the unpredictable dynamics of a spacecraft in water may pose a hazard to rescue teams and vehicle-specific rescue equipment may also need to be brought in before committing rescuers from a ship of opportunity. But just as the infrastructure that would rescue mariners in distress adapted to assist downed aviators, so will these layers of SAR capability adapt to assist space travelers in distress.

9 Conclusion

Since the early days of Vostok and Project Mercury, space programs have grown in capability. Though risks have become better understood through the lessons learned from prior missions, crewed spaceflight has always been an activity that is as firmly entrenched on the technological frontier as it is on the physical frontier. The emerging commercial space vehicles will make space much more accessible and affordable, but the crews that man these vehicles will still see the same risks as their predecessors in spaceflight and the designers of these vehicles will continue to find ways to mitigate these risks. But a landing is guaranteed every time a vehicle presses upward from its launching site, and a cleverly designed abort mechanism is only as good as the safety of the landing site it commits its crew to.

Emergency egress is no less an essential skill now as it was the day Yuri Gagarin ejected from his Vostok capsule. New capsule-based spacecraft are being designed and built as the Soyuz spacecraft and its Chinese variants will continue to fly. Suborbital spacecraft may fly over large swaths of land that may have unsuitable landing strips. More than a century after Orville Wright crashed his flying machine in Fort Myers, VA in 1908, killing his sole passenger in front of 2,000 people; Captain Chesley "Sully" Sullenberger ditched his Airbus A320-214 into the Hudson River. The 150 passengers, armed with nothing more than a 2-minute briefing, successfully egressed the aircraft without loss of life and minimal injury. Orville built his aircraft as the engineers of Vostok built their spacecraft; the technological hurdle of the new frontier justified greater risks. But as spaceflight becomes more common, the culture around it grows more risk adverse.

Today, passengers board commercial aircraft on trans-polar flights equipped with no more than a long-sleeved shirt. And though footprints have been left on the moon and international teams work together on orbiting space stations, spaceflight may never be as routine as routine as air travel has become. There are still many unknowns; vehicle designs are diverse as the designs of aircraft that populated the skies over World War One Europe. Contingency plans may always be a part of space mission planning, contingency landings may always be perceived as a very real possibility, and emergency egress operations will remain for a long time one of the most important skills that flight crews will be trained to perform in preparation of a space mission.

In this book, an insight into the history of spacecraft emergency egress systems and rescue and recovery architecture has been presented, as with an understanding of the method as to how landing and post-landing contingencies are quantified and mitigated. Egress systems and procedures have indeed matured greatly since the early days of Vostok and Mercury. But just as emergency egress considerations are intimately tied with rescue and recovery architecture design, the design of space vehicles and their operational procedures need to consider the resources of the world as a whole. In this manner, spaceflight unites our world, as the possibility of a spacecraft in distress

CONCLUSION

suddenly being in need of the resources of the sovereign lands or waters of a foreign nation fosters international goodwill.

As technologies mature and grow more fail-safe, spacecraft landings will not be the risky propositions that they have been in the past; Landing systems that will control the vehicle descent and target specific landing areas will become commonplace. But even with the high reliability we associate with commercial aviation today, emergencies requiring egress do happen. Whether or not the egress is successful often depends on the smallest considerations in vehicle design and flight crew preparation.

Appendix A: Orion Post-Landing Unassisted Egress Procedures

The following procedures were developed for the Post-Landing Orion Recovery Test Phase Two (PORT II) to validate a methodology to conduct effective unassisted egress operations from the Orion spacecraft. The initial procedures are assumed to be performed shortly after landing, followed by the 'Orion Unassisted Egress from Stable One Orientation' procedures if the spacecraft has landed in a Stable One orientation or has effectively flipped into one through successful operation of the CMUS system. Procedures detailing an egress out of the side hatch and, if sea conditions demand it, the docking hatch follows, as with the procedures for conducting a Stable Two egress out of the docking hatch in the event of a CMUS failure.

Initial Procedures

The initial procedures are assumed to be performed shortly after landing. If a hazardous environment is the cause of the unassisted egress, then it would certainly not be advisable for any crew member to open their suit visor, as once the first crew member breaks the mask seal by raising their visor, all crew members will be exposed to the environment within the capsule. Assuming that a hazardous environment does exist, the following procedures will be conducted:

- Verify capsule has landed
- Verify the internal air is free of toxic contaminants (e.g., hydrazine, ammonia, smoke)
- Open suit visor (ALL)
- Remove suit gloves (ALL)
- Attempt radio contact with ground forces to determine time to assist (CDR or PLT)

Orion Unassisted Egress from Stable One Orientation

The procedures listed in this section are performed once an emergency egress has been declared and the initial procedures have been performed. The stable-one orientation can be achieved either by landing or through a successful activation of the CMUS after a landing in a stable-two orientation. Assuming that the air suit loop is exhausted or otherwise unavailable, the egress procedure must be performed using the EBS system via the following procedures:

- Activate Emergency Breathing System (ALL)

- Powerdown / safe vehicle (CDR)
- Release seat restraint (ALL)
- Release umbilical and stow out of egress path (ALL)
- Remove/fold seat lateral supports (leave supports in place with/without pins)
- Egress seat – fold seats as necessary (fold leg pans and possibly remove seat pans)

Mission Specialists:
- Retrieve egress blanket
- Retrieve Life Raft
- Retrieve Crew Survival Kits
- Attach raft tether to CM interior attach point
- Attach crew survival kits to raft tether

Orion Unassisted Side Hatch Egress from Stable-One Orientation

The following procedures are used if there is need for an unassisted emergency egress and sea conditions are benign enough so that there would not be a sinking risk if the side hatch were opened.

- Open Side hatch (P4)
- Deploy egress blanket (P4)
- Deploy raft and kits (P4)
- Attach suit tether to raft tether, probably in order of P4, P3, P2, P1.
- Inflate LPU at side hatch (simulated)
- Egress vehicle – call time at last crewmember to cross side hatch threshold
- Ingress raft. Crew may need to pull oneself along the rope to get to the raft. Detach raft from capsule only if necessary.
- Activate radio / rescue beacons

Orion Unassisted Docking Hatch Egress from Stable-One Orientation

The following procedures are used if there is need for an unassisted emergency egress and it is deemed that opening the side hatch would be too risky due to the pitching of the vehicle and the associated risk of a rapid sinking situation.

- Retrieve docking hatch lowering device (P3)
- Attach docking hatch lowering device (P3)
- Unlock docking hatch (P3)
- Lower docking hatch (P3)
- Securely stow docking hatch (P3, time permitting)
- Deploy and attach ladder from docking hatch tunnel (P3)

Pilot:

- Climb ladder into docking hatch tunnel
- While remaining inside the docking hatch tunnel, attach suit tether to forward bay attach point
- At docking hatch interface, inflate LPU (simulated)
- Exit docking hatch tunnel to forward bay

Mission Specialist 1:
- Lift raft into docking hatch tunnel and hand up to pilot

Pilot:
- Attach raft tether to forward bay attach point

Mission Specialist 1:
- Lift egress blanket into docking hatch tunnel and hand up to pilot

Pilot:
- Attach egress blanket to forward bay attach point
- Deploy egress blanket
- Deploy raft

Mission Specialist 1:
- Lift crew survival equipment into docking hatch tunnel and hand up to pilot
- Pilot:
- Attach crew survival equipment to raft side of tether.
- Descend egress blanket into raft water
- Mission Specialist 1:
- Climb ladder into docking hatch tunnel
- While remaining inside the docking hatch tunnel, attach suit tether to forward bay attach point
- At docking hatch interface, inflate LPU (simulated)
- Exit docking hatch tunnel to forward bay
- Descend egress blanket into raft water

Mission Specialist 2:
- Climb ladder into docking hatch tunnel
- While remaining inside the docking hatch tunnel, attach suit tether to forward bay attach point
- At docking hatch interface, inflate LPU (simulated)
- Exit docking hatch tunnel to forward bay
- Descend egress blanket into water
- Commander:
- Climb ladder into docking hatch tunnel

- While remaining inside the docking hatch tunnel, attach suit tether to forward bay attach point
- At docking hatch interface, inflate LPU (simulated)
- Exit docking hatch tunnel to forward bay
- Descend egress blanket into water
- Ingress raft
- Activate radio / rescue beacons

Egress from Stable-Two Orientation if CMUS System Fails

If the Orion capsule lands in a stable-two orientation and the CMUS fails, the crew is left in a vulnerable situation. As the design of the spacecraft slightly favors a stable-one orientation, it is possible that it may orient itself that way as a result of sea motion, but sea conditions that agitated may make an unassisted an egress extremely risky proposition, but then again, so is remaining in the spacecraft with an uncontrollable hazardous environment.

If the CMUS system fails, the crew will be suspended by their seats in an inverted position. It is important that the crew extract themselves from the suspended position, as crews so suspended have a limited period of consciousness, and it may take up to three minutes to verify that the CMUS has, indeed, failed. The following set of procedures assumes that the egress initiator has occurred within the first three minutes following landing, the crew remains suspended in their seats, and it has been determined that the CMUS has failed.

- Activate Emergency Breathing System – Assume that suit loop not functional after landing, so the emergency breathing system must provide sufficient breathing air for the full egress.
- Close visors (ALL)
- Powerdown / safe vehicle (CDR)

Commander / Pilot:
- Release seat restraint
- Release umbilical and stow out of egress path
- Remove/fold seat lateral supports (leave supports in place with/without pins)
- Egress seats, but do not fold seats as they will be used as handholds / supports

Commander:
- Approach Mission Specialist 1
- Holds Mission Specialist 1's legs
- Mission Specialist 1:
- Once Commander has secured legs, release seat restraint
- Release umbilical and stow out of egress path
- Remove/fold seat lateral supports (leave supports in place with/without pins)

- Egress seats, but do not fold seats as they will be used as handholds / supports

Pilot:
- Approach Mission Specialist 2
- Holds Mission Specialist 2's legs

Mission Specialist 2:
- Once Commander has secured legs, release seat restraint
- Release umbilical and stow out of egress path
- Remove/fold seat lateral supports (leave supports in place with/without pins)
- Egress seats, but do not fold seats as they will be used as handholds / supports
- Retrieve Life Raft
- Retrieve Crew Survival Kits
- Retrieve docking hatch lowering device
- Attach docking hatch lowering device

Commander:
- Climb into docking hatch tunnel
- Activate hatch pressure equalization system
- Wait for water to stop entering the capsule
- Unlock docking hatch
- Exit the docking hatch tunnel
- Raise docking hatch into capsule
- Securely stow docking hatch
- Attach crew survival kits to raft tether

Pilot:
- Climb into docking hatch tunnel
- Hold raft tether attaching device
- Assume a head down position

Commander:
- While holding the pilot's legs, push the pilot downward into the water but do not let go

Pilot:
- Attach raft tether to forward bay attach point
- Indicate to Commander that raft is attached

Commander:
- Once indication of raft attachment has been received, assist pilot in re-entering capsule

Pilot:
- Re-enters the capsule
- Deploy raft
- Climb into the docking hatch tunnel
- Assume a head down position

Commander:
- While holding the pilot's legs, push the pilot downward into the water but do not let go

Pilot:
- Find raft tether
- Attach suit tether to raft tether
- Indicate to Commander that suit tether is attached to raft tether
- Commander:
- Once indication of suit tether attachment has been received, push pilots legs as far into water as possible

Pilot:
- Grip forward bay and pull yourself out of the docking hatch tunnel
- Ascend raft tether
- Upon reaching the surface of the ocean, inflate LPU
- Ingress raft

Mission Specialist 1:
- Climb into docking hatch tunnel
- Assume a head down position

Commander:
- While holding MS1's legs, push the MS1 downward into the water but do not let go

Mission Specialist 1:
- Find raft tether
- Attach suit tether to raft tether
- Indicate to Commander that suit tether is attached to raft tether

Commander:
- Once indication of suit tether attachment has been received, push MS1's legs as far into water as possible

Mission Specialist 1:
- Grip forward bay and pull yourself out of the docking hatch tunnel
- Ascend raft tether

- Upon reaching the surface of the ocean, inflate LPU
- Ingress raft

Mission Specialist 2:
- Climb into docking hatch tunnel
- Assume a head down position

Commander:
- While holding MS2's legs, push the MS2 downward into the water but do not let go

Mission Specialist 2:
- Find raft tether
- Attach suit tether to raft tether
- Indicate to Commander that suit tether is attached to raft tether

Commander:
- Once indication of suit tether attachment has been received, push MS2's legs as far into water as possible

Mission Specialist 2:
- Grip forward bay and pull yourself out of the docking hatch tunnel
- Ascend raft tether
- Upon reaching the surface of the ocean, inflate LPU
- Ingress raft

Commander:
- Climb into docking hatch tunnel
- Assume a head down position
- Grip forward bay and pull yourself out of the docking hatch tunnel
- Find raft tether
- Attach suit tether to raft tether
- Ascend raft tether
- Upon reaching the surface of the ocean, inflate LPU
- Ingress raft
- Detach raft from capsule (CDR)
- Activate radio / rescue beacons (CDR or PLT)

Appendix B: A Shortlist of Crewed Spaceflight Anomalies

1.   1961 July 21 - Mercury MR-4. Hatch blew after splashdown; capsule sank; astronaut barely saved before drowning.

2.   1962 February 20 - Mercury MA-6. Erroneous warning signal led ground to believe landing bag had deployed, dooming astronaut to incineration on re-entry. Re-entry was accomplished with the retro-rocket pack retained, resulting in a spectacular and unpredictable return to earth.

3.   1962 May 24 - Mercury MA-7. Excessive fuel use and pilot error led to late re-entry, and landing 300 km past the intended point. Capsule ran out of orientation fuel during re-entry.

4.   1963 May 15 - Mercury MA-9. In long duration mission, virtually all capsule systems failed. Nevertheless the astronaut was able to manually guide the spacecraft to a pinpoint landing.

5.   1963 June 16 - Vostok 6. Tereshkova did not reply during several communications sessions. To this day it is not known if she was paralyzed with fear, or if there was an equipment failure.

6.   1965 March 18 - Voskhod 2. Cosmonaut barely able to get back into air lock after world's first spacewalk. Oxygen leak flooded cabin, creating fire danger. Manual re-entry when main system failed, resulting in landing in Ural Mountains. Crew spent night in woods, surrounded by wolves, before recovery crews arrived.

7.   1965 June 3 - Gemini 4. Astronaut could barely get back into capsule after first American spacewalk. Failure of spacecraft computer resulted in high-G ballistic re-entry.

8.   1965 December 15 - Gemini 6. First launch attempt resulted in shut-down of the main engines a second after they had already ignited. The booster settled back onto the pad unsecured. The crew decided not to eject.

9.   1966 March 16 - Gemini 8. A stuck thruster aboard Gemini resulted in the crew nearly blacking out before the resulting spin could be stopped. An emergency landing in the mid-Pacific Ocean followed.

10.   1966 June 3 - Gemini 9. Could not dock with target vehicle due to jammed shroud. During spacewalk, the astronaut became exhausted and face plate fogged over. He was barely able to return to the spacecraft and close the hatch.

11.   1966 July 18 - Gemini 10. Astronaut lost grip in space walk from Gemini to Agena, tumbled head over heels at end of umbilical around Gemini.

12.   1967 April 23 - Soyuz 1. Solar panel failed to deploy. Mission aborted and successful retrofire completed. However capsule's main parachute would not jettison from container and astronaut killed on impact with earth.

13.   1968 December 21 - Apollo 8. Crew had no 'lifeboat' in case of massive failure en route to moon (as would happen later with Apollo 13).

14.   1969 January 15 - Soyuz 5. Service module failed to separate resulting in nose-first re-entry. The bolts connecting the service module to the re-entry capsule finally burned through and the capsule turned around, heat shield forward, just before the forward hatch melted. All capsule propellant was exhausted and the cosmonaut made a 9-G uncontrolled re-entry, landing hundreds of kilometres short.

15.   1969 January 15 - Soyuz 4/5. Suit hung up on attempt to exit spacecraft and flow of oxygen was shut off. Fixing this diverted the crew, resulting in no film of the world's first crew transfer between two spacecraft.

16.   1969 May 18 - Apollo 10. Incorrect switch setting led to wild gyrations when the LM ascent stage separated at an altitude of 15 km above the lunar surface. The crew regained control only two seconds before the LM would have been on an irreversible course to crash on the Moon.

17.   1970 April 11 - Apollo 13. Fuel cell tank exploded en route to the Moon, resulting in loss of all power and oxygen. Only through use of the still-attached LM as a lifeboat could the crew survive to return to earth.

18.   1970 June 1 - Soyuz 9. Head-over-heels rotation of Soyuz to conserve fuel and lack of exercise resulted in terrible condition of astronauts on return. The Soviets almost reconsidered their space station plans as a result.

19.   1971 April 23 - Soyuz 10. Hard dock with station could not be achieved. When the astronauts tried to pull away, they were stuck and could separate from the station only after repeated attempts. Toxic fumes in air supply during landing overcame one astronaut.

20.   1971 June 6 - Soyuz 11. Main telescope inoperative due to failure of cover to jettison. Fire in space station nearly resulted in emergency evacuation. Fail-safe valve opening during re-entry resulted in decompression and death of entire crew.

21.   1971 July 26 - Apollo 15. One of the three main parachutes failed, causing a hard but survivable splashdown.

22.   1973 July 28 - Skylab 3. Leaks in Apollo CSM thrusters led to preparation of a rescue mission. In the end it was decided to attempt a landing with the faulty thrusters.

23.   1975 April 5 - Soyuz 18-1. During launch third stage separation failed to occur. Crew aborted to 20 G landing in mountains near Chinese border, sliding down a slope towards a cliff until their parachute snagged on a tree.

24.   1975 July 15 - Apollo (ASTP). Crew nearly killed by toxic propellant vapors dumped into the cabin air supply during re-entry.

25.   1976 July 6 - Soyuz 21. Crew member became psychotic and mission was returned to earth from space station early. Toxic gases in station were suspected.

26.   1976 October 14 - Soyuz 23. Docking aborted due to electronics failure. Crew nearly froze to death after an emergency landing in a lake in a blizzard at -20 deg C. It took hours before the capsule could be dragged to shore.

27.   1983 September 26 - Soyuz T-10-1. Launch vehicle blew up on pad, crew rescued by launch escape tower, which pulled their capsule away at 20 G's.

28.   1985 July 29 - STS-51-F. Second engine-start pad abort of the program. First ascent abort when the center SSME shut down three minutes early due to faulty engine temperature sensors. At T+645 seconds the number one engine shut down prematurely due to a sensor problem. An abort to orbit was declared. Also experienced a blow hole through the putty in the right-hand SRM nozzle and the primary O-ring was affected by heat.

29.   1986 January 12 - STS-61-C. On a 6 January 1986 launch attempt, a temperature probe inside a propellant line broke off and went into a fluid control valve in one of the SSME's, jamming it in the open position. Had the launch not been scrubbed for other reasons, the valve probably would have caused a turbo pump engine overspeed at engine shutdown, resulting in disintegration, and loss of both nearby hydraulic systems. Columbia would have made it to orbit, but been unable to return to earth. This would have been compounded by a massive undetected loss of liquid oxygen propellant before the launch. This would have meant Columbia would have run out of propellant, not reached orbit, then lost its hydraulic systems, and then burned up on re-entry!

30.   1986 January 28 - STS-51-L. An O-ring failure in a solid rocket booster led to leaking of hot gases against the external tank. The resulting explosion killed the seven member crew.

31.   1997 April 4 - STS-83. Orbiter recalled to earth after three days of flight when one of three fuel cells failed. Mission reflown as STS-94.

32.   1999 July 23 - STS-93. A repair pin in an SSME 3's combustion chamber came loose, punching a hole in the nozzle cooling jacket. Hydrogen fuel leaked out during the ascent, resulting in Columbia running out of gas and ending up in an orbit seven miles lower than planned. At the same time a short circuit had disabled two of the SSME's engine controllers at five seconds into the flight, so the ascent was made on remaining backup controller, with no remaining backup.

33.   2002 November 24 - ISS EO-6. OMS valve stuck on shuttle during ascent, requiring orbital maneuvers with the single remaining engine. On return to earth aboard Soyuz guidance failed and a ballistic entry subjected the crew to over 8 G's and a landing 460 km short of the planned location.

34.   2003 January 16 - STS-107. Crew perished when shuttle broke up during re-entry. Cause was damage to a leading-edge RCC from foam breaking off of external tank bipod strut.

35.   2004 June 21 – Spaceship One Flight 15P. Spacecraft rolled 90 degrees right and left at motor ignition; attitude control lost at engine shut down; engine fairing collapsed.

36.   2007 April 7 - ISS EO-15. The re-entry burn began at 09:47 and was normal. But afterwards, due to failure of an explosive bolt, the Soyuz service module remained connected to the re-entry capsule. The Soyuz tumbled, and then began re-entry with the forward hatch taking the re-entry heating, until the connecting strut burned through. The Soyuz the righted itself with the heat shield taking the heating, but defaulted to an 8.6 G ballistic re-entry, landing 340 km short of the aim point at 10:36 GMT. Improved procedures after the ballistic re-entry of Soyuz TMA-1 meant a helicopter recovery crew reached the capsule only 20 minutes after thump down. However the true nature of the failure was concealed from the world until the same thing happened on Soyuz TMA-11.

37.   2007 October 10 - ISS EO-16. Following the de-orbit burn at 07:40 GMT the aft service module of the Soyuz failed to separate and the spacecraft began re-entry in a reversed position, with the forward hatch taking the initial re-entry heating. As was the case with Soyuz 5 in 1970, the connections with the service module finally melted away, and the freed capsule righted itself aerodynamically with the heat shield taking the brunt of the re-entry heating. However the crew experienced a rough ride, a ballistic re-entry of over 8 G's force, smoke in the cabin, a failure of the soft landing system, and a very hard landing. They landed 470 km short of the target point at 50 deg 31" N, 61 deg 7" E at 08:29 GMT. A small grass fire was started at the landing point and the injured crew had to be helped from the capsule by passers-by. Malenchenko and Whitson suffered no permanent injury, but Yi was hit by Whitson's personal effects bag on impact and required physical therapy for neck and spine injuries.

References

1   35 años del Soyuz 18-1 (7 April 2010), Retrieved 1 August 2012 from Daniel Marin personal blog: http://danielmarin.blogspot.com/

2   AMVER Home page, U.S. Coast Guard, Retrieved 9 February 2012 from www.amver.com

3   *Apollo Hatch Redesign, A Matter of Urgency'*, Historic Space Systems, Info Sheet, Issue 2 – Apollo, December 1996, Issue 2 – Apollo

4   *Apollo Operations Handbook: Spacecraft 012'*, SID 65-1317, *North American Aviation, Inc.*

5   Apollo Operations Handbook Block II Spacecraft', Vol. 1' NASA Document NAS 9-150, 15 April 1969.

6   *Astronaut Training for the Mercury Flights*. Retrieved 1 August 2012 from Angelfire website: http://www.angelfire.com/space2/sp425/5.html

7   Barrett, M.R. and Pool, S.L., *'Principles of Clinical Medicine for Space Flight'*, 1st edition, May 1, 2008. ISBN-10: 0387988424.

8   *Bitter Aftertaste of Glory*, M. Rebrov, 'Krasnaya Zvezda', Sep 9, 1994, p.2.

9   Brinkley, J.W., 'Personal Protection Concepts for Advanced Escape Systems Design', Air Force Aerospace Medical Research Laboratory, Wright-Patterson Air Force Base.

10 Burrough, Bryan (1998), *Dragonfly: NASA and the Crisis Aboard Mir*, HarperCollins, p. 185, ISBN 0-88730-783-3.

11 Caires, S. and Swail, V., 'Global Wave Climate Trend and Variability Analysis', Meteorological Service of Canada and Royal Netherlands Meteorological Institute, Meteorological Service of Canada, Toronto, Ontario, Canada.

12 Chandler, M., Personal interview, private communication, September 2009.

13 *Civil Air Search and Rescue Association (CASARA)*, Retrieved 1 August 2012 from the Civil Air Search and Rescue Association website: http://www.casara.ca/

14 Clancy, H., 'NASA Program Apollo Working Paper No. 1348, Evaluation of the Crew – Command Module Postlanding Interface', 7 July 1969.

15 Clancy, Harold J.; and Dailey, Reed M.: Crew Egress Procedures for Apollo Block I Command Module at Sea. NASA Program Apollo Working Paper No. 1213, 1966.

16 *Cold War Leftovers.* Retrieved 1 August 2012 from Duotone website: http://www.duotone.com/coldwar/spacerace/index.html

17 Crew Egress Procedures for Apollo Block 1 Command Module at Sea', NASA Program Apollo Working Paper No. 1213. December 7, 1966.

18 Crew Health Operations Concept for Constellation Missions Space Life Sciences Doctorate, NASA, JSC-63566, August 2006.

19 Dennis R. Jenkins: 'Space Shuttle - The History of Developing the National Space Transportation System', Dennis R. Jenkins Publishing 1999, Page 272, ISBN 0963397443

20 *Descenders & Belay Devices*, Retrieved 1 August 2012 from Rope and Rescue website: http://www.ropeandrescue.com/descenders/

21 Diller, G., *Space Shuttle Weather Launch Commit Criteria and KSC End of Mission Weather Landing Criteria* (January 2003). Retrieved 1 August 2012 from NASA website: http://www.nasa.gov/centers/kennedy/news/releases/2003/release-20030128.html

22 Do, S. and de Weck, O., 'An Airbag-Based Impacr Attenuation System for the Orion Crew Exploration Vehicle', thesis submitted to the Department of Aeronautics and Astronautics, Massachusetts Institute of Technology, February 2011.

23 Encyclopedia Astronautica. *HL-20*. Retrieved 1 August 2012, from Encyclopedia Astronautica website: http://www.astronautix.com/craft/hl20.htm

24 *Everybody Out!* Retrieved 1 August 2012 from NASA website: http://www.nasa.gov/audience/foreducators/k-4/features/F_Everybody_Out_prt.htm

25 Extended Duration Orbiter Medical Project (EDOMP), NASA/SP–1999-534.

26 Ferranti, L., T. N. Palmer, F. Molteni, E. Klinker, 1990: Tropical-Extratropical Interaction Associated with the 30-60 Day Oscillation and Its Impact on Medium and Extended Range Prediction. Journal of the Atmospheric Sciences:Vol. 47, No. 18, pp. 2177-2199.

27 Flatau, Maria, Piotr J. Flatau, Patricia Phoebus, Pearn P. Niiler, 1997: The Feedback between Equatorial Convection and Local Radiative and Evaporative Processes: The Implications for Intraseasonal Oscillations. Journal of the Atmospheric Sciences: Vol. 54, No. 19, pp. 2373-2386.

28 Gagarin, Yu., Doklad Tov. Gagarina Ya. A. ot 13 aprelya 1961 ha zasedanii Gosudarsvennoi komissii mosle kosmicheskogo poleta, (quoted from a published version in Aero, No. 1, p. 10)

29 Gerasimov, A.,' Correspondent at landing site, Moscow Central TV, Oct 10, 19:00 GMT.

30 Grahn, S., 'Soyuz emergency landing zones - the "Ugol Pasadki" story, retrieved 1 August 2012 from svengrahn.pp.se: http://www.svengrahn.pp.se/histind/Ugol/Ugol.html

31 Hall, R., and Shayler, D., 2001, "The Rocket Men: Vostok & Voskhod, the first Soviet manned spaceflights," Springer; Published in association with Praxis, London ; New York; Chichester England, pp. 326.

32 Hall, Rex; David Shayler (2003). *Soyuz: A Universal Spacecraft*. Springer. pp. 195–196. ISBN 1-85233-657-9.

33 Hyle, C. T. et al., "Apollo Experience Report - Abort Planning", NASA Technical Report TN D-6847, June 1972.

34 Jones, E., '*Apollo 13 Image Library*' (10 February 2012). Retrieved from NASA website: http://www.hq.nasa.gov/alsj/a13/images13.html

35 *Kazakhstan to acquire 20 Eurocopter EC725 helicopters*, Retrieved 1 August 2012 from Brahmand website: http://brahmand.com/news/Kazakhstan-to-acquire-20-Eurocopter-EC725-helicopters/9514/1/24.html

36 Knaff, P.R., *On-pad Crew Safety: Modifications to Operations and Hardware to Improve Crew Safety Times* – Case 320. NASA Document CR-153840, 3 March 1967.

37 Kuznetsky, M. I., Gagarin na Kosmodrome Baikonur, Vladi, Krasnoznamenk, 2001

38 Landing, postlanding, and SRB retrieval, NSTS Shuttle Reference Manual (1988), Retrieved 1 August 2012 from http://science.ksc.nasa.gov/shuttle/technology/sts-newsref/stsover-landing.html

39 Linde, M., 'Soyuz Experiences/Contingencies', 24 June 2008.

40 Mader, T.H. et al., 'Optic Disc Edema, Globe Flattening, Choroidal Folds, and Hyperopic Shifts Observed in Astronauts after Long-duration Space Flight", Ophthalmology Volume 118, Issue 10 , Pages 2058-2069, October 2011.

41 Malik P.W. and Souris G.A., 'NASA CR-1106 Technical Report, 'Project Gemini: A Technical Summary', June 1968.

42 *March to Moon Hinges on Apollo 7*', Spartanburg Herald, p32, 10 October 1968.

43 Mission Training Program for the Apollo Lunar Landing Missions, MSC Internal Note No. MSC-CF-D-6828, Manned Spaceflight Center, December 20, 1968.

44 Mission Training Plan for Gemini IX Flight Crew", NASA Program Gemini Working Paper No, 5047, February 14, 1966.

45 MSC Internal Note:MSC-CF-D-68-10, Apollo Egress Training Plan, Mission Training Section, Mission Operations Branch, Flight Crew Support Division, February 20, 1968.

46 *NASA Ejection Seats: Gemini Spacecraft*, Retrieved 1 August 2012 from The Ejection Site website: http://www.ejectionsite.com/gemini.htm

47 National Search and Rescue Manual, Australian National Search and Rescue Council, Revision No: 11, 20 Jul 2011.

48 Oberg, J., 'Soyuz Landing Historical Reliability Study', March 19, 1997. Retrieved 1 August 2012 from JamesOberg.com: http://www.jamesoberg.com/soyuz.html

49 Oberg, J., 'Internal NASA Documents Give Clues to Scary Soyuz Return Flight', IEEE Spectrum, May 2008. http://spectrum.ieee.org/aerospace/space-flight/internal-nasa-documents-give-clues-to-scary-soyuz-return-flight/0

50 Paul T. Chaput, P.T., 'Crew Egress Procedures Developed During the Qualification Test Program for the Gemini Spacecraft at-sea Operations', NASA Program Gemini Working Paper No. 5015, August 26, 1964.

51 Petty, J.I., 'Space Shuttle Aborts', retrieved 1 August 2012 from http://spaceflight.nasa.gov/shuttle/reference/shutref/sts/aborts/index.html

52 *Pitching Science: Engineers who track baseballs catch insights into the game*, Retrieved 1 August 2012 from Science News Online website: http://www.phschool.com/science/science_news/articles/pitching_science.html

53 Poletaeva, V., Shagnuvshie k Zvezdam, (In Russian) Samara, TsSKB Progress 2006

54 Project Gemini familiarization manual, McDonnell, SEDR 300 Volume 1, 30 September 1965.

55 Program Gemini Working Paper No. 5027, 'Mission Training Plan for Third Manned Gemini Flight Crew', NASA, June 21, 1965.

56 Project Mercury Capsule Flight Operations Manual, Capsule 7 (MR-3), SEDR109-7, 15 August 1960.

57 Project Mercury Capsule Flight Operations Manual, Capsule 13 (MR-3), SEDR109-13, 1 September 1961.

58 *Project Mercury* by Spacecraft Films (Disc 5). Retrieved 1 August 2012 from Collect Space website: http://www.collectspace.com/ubb/Forum29/HTML/001313.html

59 Rea, Jeremy R.; Putnam, Zachary R. (20–23 August 2007). "A Comparison of Two Orion Skip Entry Guidance Algorithms" (PDF). AIAA Guidance, Navigation and Control Conference and Exhibit. Hilton Head, South Carolina.

60 Reimuller, J.D. and Rhodes, A.C., PORT Phase II: Full Unassisted Body Analog Recovery: 8 July 2008.

61 Reimuller, J.D. and Rhodes, A.C., Post Landing Orion Recovery Test (PORT) Phase Two Update: Assisted and Unassisted Crew Egress', 29 October 2008.

62 Reimuller, J.D. et al., 'Launch Abort Landing Conditions Analysis', NASA summary report, 12 Nov 2010.

63 Reimuller, J.D., 'AMVER Response Time Distributions exceeding 24 hours', 8 December 2008.

64 Reimuller, J.D., 'Recovery Time Distributions for Ascent Abort Scenarios', 29 February 2008.

65 Reimuller, J.D., et al., 'Post-Landing Emergency Crew Egress GMO SIG-09-1028 TDS IDAC-5 Study Results', 02 November 2009.

66 Reimuller, J.D., Recovery Options in Excessive Sea Conditions and Implications to Crew Risk 30 March 2010.

67 Rhodes, A.C. et al., 'PORT 2 Emergency Egress Procedures', October 2009.

68 *Russia thriving again on the final frontier.* Retrieved 1 August 2012 from MSNBC website: http://www.msnbc.msn.com/id/9509254/ns/technology_and_science-space/#.UGI3GVHF0lQ

69 Seamans, R.C. Jr., 'Project Apollo: The Tough Decisions", Monographs in Aerospace History No. 37, NASA Document SP-2005-4537, 2005.

70 *Search and Rescue*, Retrieved 1 August 2012 from Royal Canadian Air Force website: http://www.rcaf-arc.forces.gc.ca/v2/page-eng.asp?id=17

71 Shayler, D., 2001, "Gemini: Steps to the Moon," Springer; Published in association with Praxis Pub., London ; New York; Chichester England, pp. 433.

72 Shayler, D., 2009, "Space rescue: ensuring the safety of manned spaceflight," Springer; Published in association with Praxis, Berlin ; New York; Chichester, UK, pp. 356.

73 Siddiqi, A.A., 2000, "Challenge to Apollo: The Soviet Union and the Space Race, 1945-1974," National Aeronautics and Space Administration, Washington D.C., pp. 1011.

74 Slayton, D. (1961). Pilot Training and Preflight Preparation. In Proceedings of a Conference on Results of the First U.S. Manned Suborbital Space Flight (pp. 53-60). Washington, DC: NASA. http://www.angelfire.com/space2/sp425/5.html

75 *Slidewire baskets slide into history*, Retrieved 1 August 2012 from Collecct Space website: http://www.collectspace.com/ubb/Forum30/HTML/001111.html

76 Slipchenko, S.,' Interview with crewmember suggests high winds may have contributed to impact severity.', TV report at the landing site, Ostankino TV, Aug 10, 17:00 GMT.

77 Somers, J. et al., 'Occupant Protection Project', NASA Document, October 2010

78 Soyuz-23, Lands On A Frozen Lake, Retrieved 1 August 2012, from Video Cosmos website: http://www.videocosmos.com/soyuz23.shtm Soyuz TM (Soyuz TMA)

79 Crew Extraction and Medical Support at a Contingency Landing Site', retrieved 1 August 2012 from Spaceref.com: http://www.spaceref.com/iss/soyuz/SCLSaC.edit.pdf

80 Space Shuttle Abort Modes. Retrieved 1 August 2012 from Aerospace Web website: http://www.aerospaceweb.org/question/spacecraft/q0278.shtml

81 Space Shuttle Safety Review', Audit Report JS-96-003, Johnson Space Center June 28, 1996.

82 Space Shuttle Weather Launch Commit Criteria and KSC End of Mission Weather Landing Criteria', National Aeronautics and Space Administration, Feb. 5, 2010.

83 Sterl, A. and Cairnes, S.,'Climatology, Variability and Extrema of Ocean Waves - The Web-based KNMI/ERA-4 Wave Atlas', Submitted to the International Journal of Climatology, March 23, 2004.

84 The WAM Model – A Third Generation Ocean Wave Prediction Model', The Wamdi Group, Dec 1988.

85 Training for Gemini: water egress and survival training', Retrieved 1 August 2012 from the Race for Space website: http://www.raceforspace.co.uk/page1/page9/files/CEE6432F-RfS_04_PRINT_lr%206.pdf

86 *Transatlantic Landing Abort*, Retrieved 1 August 2012 from NASA website: http://spaceflight.nasa.gov/shuttle/reference/shutref/sts/aborts/tal.html

87 Trujillo, K., 'Docking Hatch Egress Options Review', June 8, 2009

88 Trujillo, K., 'Egress Operations and Time Estimates' 8 October 2009.

89 United States National Search and Rescue Supplement to the International Aeronautical and Maritime Search and Rescue Manual", National Search and Rescue Committee, Washington DC. May 2000

90 USA009026, 'Crew Escape Systems 21002', 17 January, 2005.

91 Voas, R., 'Project Mercury: A Description of the Astronaut's Task in Project Mercury', NASA Space Task Group, Presented at the Fourth Annual Meeting of the Human Factors Society, Sept 14, 1960.

92 *Voskhod*, Retrieved 1 August 2012 from University of Oregon website: http://abyss.uoregon.edu/~js/space/lectures/lec10.html

93 Walker, T., 'CMUS Up-righting Time', 21 October 2009.

94   Walkover, L.J., et al., 'The Apollo Command Module Side Access Hatch System", JPL Technical Memorandum 33-425.

95   Wei Luty, 'Brief Summary of Deconditioned Crew Protection', May 08, 2008.

96   White, R.D., *Apollo Experience Report – Command Module Uprighting System*, NASA Document TN D-7081. March 1973

97   Woods, D., and Brandt, T., 2009, "Crew Couches," NASA History Office, [Online], URL: http://history.nasa.gov/ap16fj/02system_couches.htm [Cited: October 10th, 2010].

98   Woods, W.D., 2008, "How Apollo flew to the Moon," Springer Verlag; published in association with Praxis Publishing, New York; Chichester, U.K., pp. 412.

About the Author

Dr. Jason Reimuller is an experienced system engineer and project manager. He is currently the Executive Director the International Institute for Astronautical Sciences, and President of Integrated Spaceflight Services. Jason has led studies that involved abort scenarios, launch availability analyses, rescue and recovery trade studies, and contingency crew egress testing in support of NASA's Constellation Program.

Jason holds a Ph.D. in Aerospace Engineering Sciences from the University of Colorado in Boulder. He also holds a M.S. degree in Physics from San Francisco State University, a M.S. degree in Aviation Systems from the University of Tennessee, a M.S. Degree in Aerospace Engineering from the University of Colorado, and a B.S. degree in Aerospace Engineering from the Florida Institute of Technology. He is an instrument-rated commercial pilot who has used research aircraft to obtain imagery of noctilucent clouds synchronized with satellite overpasses in Northern Canada and will soon use suborbital spacecraft to further study these clouds and their roles in the upper atmosphere.

Jason currently resides in Boulder, Colorado and travels frequently to Greenland and Guatemala. When not working on astronautics projects, Jason enjoys playing rugby, playing guitar, and skiing. Jason is also a NAUI SCUBA Divemaster.

INDEX

X

Y

Z

www.ingramcontent.com/pod-product-compliance
Lightning Source LLC
Chambersburg PA
CBHW081342190326

41458CB00018B/6079